NAKED GENES

NAKED GENES

REINVENTING THE HUMAN IN THE MOLECULAR AGE

HELGA NOWOTNY AND GIUSEPPE TESTA

TRANSLATED BY MITCH COHEN

THE MIT PRESS

CAMBRIDGE, MASSACHUSETTS

LONDON, ENGLAND

Die gläsernen Gene. Die Erfindung des Individuums im molekularen
Zeitalter © Suhrkamp Verlag Frankfurt am Main 2009.

For information about special quantity discounts, please email special_
sales@mitpress.mit.edu.

This book was set in Engravers Gothic and Bembo by the MIT Press.
Printed and bound in the United States of America.
With thanks to the Riksbanken Jubileumsfond in Stockholm, Sweden,
for funding the translation from German.

Library of Congress Cataloging-in-Publication Data

Nowotny, Helga.
[Gläsernen Gene. English]
Naked genes : reinventing the human in the molecular age / Helga
Nowotny and Giuseppe Testa ; translated by Mitch Cohen.
    p.   cm.
Includes bibliographical references and index.
ISBN 978-0-262-01493-9 (hardcover : alk. paper)
1. Molecular biology—Social aspects. I. Testa, Giuseppe. II. Title.
QH506.N6913   2011   303.48'3—dc22   2010017756

10   9   8   7   6   5   4   3   2   1

CONTENTS

PREFACE

The term *society* is precisely as useful for sociologists as the term *life* is for biologists: an empty, regulatory term that describes the final, asymptotic goal of their research.

—Raymond Boudon, 2007

Terms like *society*, *life*, *gene*, and *individual* are currently being redefined. Driven by the forces of globalization, the content and attributed meaning of both scientific work in laboratories and the diverse experiences of everyday life are rapidly changing. New forms of scientific and social life are arising and are offering new options in the molecular age.

Although we come from different scientific backgrounds, we both are interested in the possibilities and far-reaching societal effects associated with biotechnology, especially through the developments in postgenomics and synthetic biology. These are unsettling and yet also reflect a striving for human enhancement. The gene has been made visible, the old persists in the face of radical discontinuities, and law and bioethics play roles in the governance of life. These all bring to light the contours of a latent, disputed future that must be shaped. What is special about this future is its double shape: the scientific side is mirrored in the societal and vice versa.

This book was made possible by preconditions that we do not take for granted. Through the foresight and generosity of Branco Weiss, younger researchers from the life sciences are able to cross disciplinary boundaries in the framework of the fellowship program called "Society in Science." Through this program, we began our collaboration, and we continued in intense talks in Vienna, Milan, and Bonassola. Two institutions supported our work by offering short-term guest residencies—the Wissenschaftskolleg zu Berlin and the Fondation Maison des Sciences de l'Homme in Paris. Hanni Geiser was, as always, an intelligent and attentive reader of our text; we gladly made use of her critical remarks.

Our sincere thanks are owed to all who contributed to creating these conditions.

Helga Nowotny and Giuseppe Testa
Vienna and Milan, Spring 2010

I   THE VISIBILITY OF THE GENETIC FUTURE

THEME I   To a much greater degree than generally realized, the often sensational achievements in the life sciences stand in continuity and commonality with old practices tested in social living. What is playing out today in the focus of public controversies has been previously thought through, dreamed of in myths, or prefigured in the joint history of the domestication of plants and animals. But discontinuities in the life sciences put pressure on longstanding social arrangements. A defining feature of the molecular life sciences—equally as important as (if not more important than) their well-recognized ability to reshape organisms and bodies—is that they make things visible that could not previously be seen. Extracted from their original contexts (and placed into new ones), these things tend to acquire, precisely through their newly found visibility, an essential status of their own. They are thus—falsely—seen as agents that can act on their own. This thesis is elucidated in examples from two areas in which the newly achieved visibility and mobility play large roles and in which the tension between continuity and discontinuity becomes salient—assisted reproduction technologies (ART) and the striving to enhance achievement with its controversial but trailblazing potential.

MAKING THINGS VISIBLE

With the slogan "Sense and Simplicity," the Philips company's global advertising campaign promotes a number of devices intended to simplify daily life. A clever design shows a minimalist white box with the simple label: "Technology should be as simple as the box it comes in." This creates the impression that a device finally fulfills the dream of immediate access. Among Philips's new gadgets is a three-dimensional ultrasound scanner for prenatal diagnosis. In the advertisement, first the viewer sees the conventional two-dimensional image of a fetus. A landscape in shades of gray is shown; only the trained eye of a gynecologist can make out its relevant contours. The translation follows in the second image. The relevant contours are marked with a continuous white line running through the gray landscape. Now the shape of a thumb-sucking fetus is recognizable. The picture thus corresponds with what ought to become visible in a routine obstetrics exam: the physician explains to the excited parents the shape of the fetus, which is initially visible on the monitor as a moving, gray pattern. Then the third image delivers the advertisement's promise—simplicity. This is the 3D ultrasound scan that uses an algorithm to transform the meaningless segments of the gray surfaces into the familiar 3D image of a baby sucking its thumb. The image "speaks" for itself; the baby's head is now recognizable even for laypeople.

Philips's advertising specialists probably were not thinking of Diderot when they explained their product's attractiveness by its ability to make things visible. Diderot's *D'Alembert's Dream* (1769) records a conversation between Diderot and his friend D'Alembert about matter and the characteristics of living things. The author demonstrates his viewpoint in an initially

paradoxical-seeming comparison between a piano endowed with
sentience and memory and the development of the egg:

> Do you see this egg? With this you can topple every theological theory,
> every church or temple in the world. What is it, this egg, before the seed
> is introduced into it? An insentient mass. And after the seed has been
> introduced into it? What is it then? An insentient mass. For what is the
> seed itself other than a crude and inanimate fluid? How is this mass to
> make a transition to a different structure, to sentience, to life? Through
> heat. And what will produce that heat in it? Motion . . . [1]

The point here is not to revive the debate between vitalism and
materialism but merely to note that, at the conclusion of the debate, Diderot considers it enough to show things as they are. For
him, seeing an egg as such disproves all schools of theology and
all the temples in the world.

Two and a half centuries later, our abilities to visualize things
have increased. We no longer see merely an egg but also see
what goes on inside it. In addition to the microscope, a number
of other instruments analytically separate and make visible the
building blocks of an egg and other components of biological
matter—genes, proteins, and intracellular membranes and com-
partments. Life is subdivided into its organizational units. Our
molecular gaze makes this fragmentation possible. And this step
seems to follow logically from the dissection of corpses in the
Renaissance, with genome browsers replacing anatomical the-
aters as if to show that only the depth and resolution of our gaze
have changed. Today our gaze pierces genes instead of organs.

But if our view today has grown sharper and deeper, why
don't things and their contexts become simpler? If observing an
egg in its motion and warmth is enough to topple every theo-
logian on earth, why has seeing the molecular mechanisms that

underlie that motion and warmth led to such heated controversies over the life sciences? Far from being disproved, the schools of theology and the temples associated with them have themselves adopted the molecular glance. They too interpret and evaluate the elementary building blocks of the molecular age. In public discourse, they have become powerful stakeholders that codetermine what should happen with the molecules that determine the organic whole. Secular ideas of morality are also shaken by what the molecular glance reveals. Classical secular tropes—human rights, human dignity, human equality—are now projected in an unforeseen way onto the fragments of life that have now become visible. As Alex Mauron rightly notes: "The genome has become the secular equivalent of the soul."[2] So what is wrong with the molecular visibility of life, which has become such a prominent characteristic of our time? Where does the alarm come from?

WHAT EFFECT DOES LIFE'S NEW VISIBILITY HAVE?

The molecular view does not simplify things but complicates them and leads to controversies because it enables interventions in life on a scale that did not exist in the past. What the external view of an egg enables people to do with it is very limited. By contrast, revealing its internal functioning uncovers almost endless possibilities for manipulating its functions and putting them together in a new architecture of life. It is no coincidence that cloning—the somatic transfer of cell nuclei—has become the icon of the potential to alter life. Replacing the cell nucleus with the genome taken from another cell of the same or a different species is a vivid illustration of what the molecular glance can do with the knowledge and technology it comprises. It also shows that the familiar distinctions—between knowledge and application, between science and technology—are outdated. Under the he-gemony of the molecular glance, knowledge has become action. Today the fact is that *understanding life means changing life*.

The molecular life sciences' glance from within has replaced the external view—the famous "view from nowhere."[3] The lat-ter postulated the ideal of nature as an object existing "out there" as a collection of truths to which science needed only hold up its innocent mirror. It was possible in Diderot's lifetime to see an egg without doing anything with it, but this strict separation has as good as vanished today. The anthropologist Paul Rabinow comments on the sequencing of the human genome:

The object to be known—the human genome—will be known in such a way that it can be changed. This dimension is thoroughly modern; one could even say that it instantiates the definition of modern rational-ity. Representing and intervening, knowledge and power, understanding and reform, are built in, from the start, as simultaneous goals and means. [4]

All projects that seek to understand life aim at changing it, which is hardly astonishing considering the rationality that drives these developments. For example, the examination of the earliest phas-es of embryonic development can be carried out only by means of the same laboratory protocols and instruments that enable the later modification of these very phases. What sounds like an epis-temic tautology is in reality the engine of scientific-technological development.

In the face of this fact, it is no wonder that the molecular glance has drawn the attention of theologians and their temples, whether religious or secular. The project of the Enlightenment planned to let nature finally speak for itself and to register its voice encyclopedically. It was based on an unspoken assump-tion about the moral authority of nature, an assumption found in various forms and nuances in many cultures. Diderot's claim was implicitly based on the idea that, by viewing nature, we could unconsciously derive norms for dealing with it. The sight of the egg (the question of *what is*) was to ground the normative discourse about values and meaning, translated one to one into the established territory of temples and theologians (the question of *what ought to be done*). But if what is "given by nature" becomes predicated on the gaze of molecular biology with the options for intervention it implies, then what is natural is from then on subject to the contingency of such interventions. *What is* multiplies into numerous options of *what ought to be* (or could be). In this sense, on the molecular level what is natural is becoming a substantially political issue.

But are we really in the process of crossing an anthropological threshold that we should approach only with the greatest caution, if at all? Do we really stand on the brink of an epochal rupture in the history of humankind, a point of no return? The entry into the molecular age, which like all periodizations can never be exactly pinned down, in no way extinguishes all previous ages. On the contrary, the new enters into new configurations with the old. The feeling of standing on the threshold of a new age is initially nothing more than that—a feeling, an inkling of the change that often begins when new concepts are introduced or new phenomena are first recognized (and named) as such. If the viewpoint and the scientific understanding of what a gene and what genomics are have dramatically changed in the last few years, then perhaps it is not surprising that these new interpretations have not yet percolated into public understanding. But vice versa, anchored in everyday knowledge and experience are practices (and the memory of them) that can ease the identification of connections and continuities between the old and the new. We interpret also in everyday life, and there too the interpretations of the forms of life and of living together are constantly changing.

AN INTERRUPTED CONTINUITY: ASSISTED REPRODUCTION

Assisted reproduction is the area of biomedicine in which the latter's potential was first actualized, and to a degree it reveals what will be possible in the future. A rough overview of the last three decades in the history of assisted reproductive technologies (ART) creates the impression that the expansion of the possibilities to conceive children has in fact led to a conspicuous discontinuity in the reproduction of life. Women after menopause, same-sex couples, single people (usually women), and even the deceased (only men so far) are examples of individuals or categories of people for whom the conception of offspring was simply out of the question before the availability of ART. The most conspicuous and far-reaching effect of this discontinuity is the desynchronization of the structuring of kinship and family relations, which is accompanied by a redistribution of the parental roles and their modalities. In the words of Marilyn Strathern, relatives are always a surprise.[5] This is all the truer when ART plays a role.

Yet what aspect of the desynchronization of the conventional family can really be traced to ART? Isn't this desynchronization due more to the unparalleled extension of human life expectancy and other improvements in living conditions, so that people have more time and opportunity to found patchwork families in various phases of life? This is not entirely true because ART undeniably entails various degrees and forms of desynchronization, some of which are inconceivable in the context of natural reproduction. The legal challenge won by Diane Blood marks a case in point. She had sperm taken from her comatose husband before his death, used it to carry out in vitro fertilization (IVF), and demanded from the British courts permission to implant the embryo. In 1998, this act seemed the most extreme example of an ART-induced desynchronization of a genealogical relationship. What could be more asynchronous than procreating new life after one's own death? We return to this case later. Here it suffices to note that the court decision established posthumous fatherhood as a legitimate family relationship and thereby granted every man (in Britain) the right to decide during his life whether he would like to procreate offspring after his death by means of ART.

Ten years later, most people regard cloning as a much more extreme scenario than IVF is. Many of the objections to the further development of reproductive cloning have to do with the suspected changes that such a practice would bring about in the genealogy that we are accustomed to—grandparents, parents, children. One frequently cited argument is that a clone would be the child and simultaneously the genetic twin of his cloned parent and thus the genetic child of its grandparents. Wouldn't this be an even less endurable degree of desynchronized confusion than what ART already causes today? It is not our aim here to provide another contribution to the monumental debate on the morality of human reproductive cloning. We are interested, instead, in the question of why cloning is perceived as such a radical challenge to our reproductive habits. In other words, how did clones become twins?

HOW CLONES BECAME TWINS

When the news of the cloned sheep Dolly went around the world in 1997, a Swedish government minister called his science adviser to vent his outrage and to demand an immediate ban on human cloning. The adviser asked him whether he had ever encountered cloned people, which the minister emphatically denied. Only after the adviser explained that monozygotic twins are "clones" did the minister calm down and stop demanding a new law.[6] There are important things in common between clones and twins, of course, but there are also differences. Clones would not fulfill three of the criteria defining twins—simultaneous conception, common prenatal environment, and the circumstance of being born together. Thus, clones cannot be twins.

A similar agitation rules the public discourse. It too is permeated by a diffuse form of genetic essentialism that sees in the genome the secular equivalent of the "soul." Scott Gilbert, one of the fathers of developmental biology, half-jokingly suggested that when meeting a stranger we could from now on take out of our pocket a CD with our genome on it instead of a visiting card. If we take seriously this conflation of genomic and personal identity, then it seems disturbing that someone else should share our same genome. But if we recall our common experience with twins, then we see two (or more) people who came into the world in the same birth process, quite apart from whether they resemble each other. In many languages, the word used for twins means nothing more than that they were born by the same mother at the same time. Indeed, another adjective, monozygotic or dizygotic, is needed to clarify whether these two people share not only their mother but also the same genome. Today we also know that dizygotic twins and about a third of all monozygotic twins do not share the placenta and the chorion in the womb and thus develop in different hormonal and endocrinological uterine environments, which may influence their later lives.

But the argument (which has become a commonplace) that a clone is merely a delayed twin shows a deeper-seated, substantial shift in our stance toward human experience—that we attribute much more importance to the supposed genetic essence than to the context in which this genetic essence develops. Incidentally, a clone would also differ genetically from its parent anyway because the clone's mitochondrial DNA comes from the egg cell of the donor (and experiments with mice have shown that mitochondrial DNA can have effects even on cognitive abilities). On top of that, the clone would in any case develop within a completely different intrauterine environment and thus encounter a host of different causal factors that mold the genotype into phenotype during embryogenesis. And it goes without saying that the environment after birth would also be different—different parents, different relationships, and a different point in time. A clone can be regarded as a twin of its parent (with whom it does not share any of its temporal experiences—the sharing of which has defined twins throughout human history) only in one sense— *that of the genome.*

Significantly enough, empirical surveys with twins have shown that they do not trace, much less reduce, their similarity and their close relationship with each other to their identical genome. Even less do they see their identity as being endangered by the fact of sharing the same genome. On the contrary, they have a positive attitude toward being an identical twin and, incidentally, have much less fear of human cloning than the average person does.[7]

THE GENETICIZATION OF CONCEPTION[8]

The most common example of ART is the donation of gametes (sperm and eggs, the haploid reproductive cells), which creates a triangular (or quadrangular) relationship between an infertile

couple and one (or two) gamete donors. This practice is still forbidden in some countries, even in the Western world. The most
conspicuous example of altered role distribution in reproduction
is surrogate motherhood, which the opponents of ART often
attack as the epitome of reproductive chaos. But a closer look
at parenting practices that have existed throughout history casts
a different light on this. In fact, even if surrogate motherhood
cannot be compared, for example, to the function that a wet
nurse had in many societies, the latter practice does show that
the transfer of the mother-child relationship to another woman
was definitely not regarded as an attack on the family. The up-
bringing and education of the children, too, were carried out
not only by the biological parents but also by a number of other
persons whose influence on the future development of the child
was probably at least as great as that of his or her intrauterine environment and hence of today's surrogate mother.

Additionally, all earlier societies dealt with children who
were not biological offspring. Social arrangements like adoption,
which existed in all cultures, testify to the ability to capture the
nuances of parenthood in a legal construction that synthesizes the
manifold influences on a child's life. Children born to unmarried
parents might have grown up with relatives or strangers. The split
between genetic and social parenthood that is often lamented as
characterizing ART has thus always existed in history—all the
more so since those were times of extended families, high rates of
child mortality and of death in childbirth, widespread patriarchy,
worries about family honor, and obsessive regulations of succes-
sion, whose economic value could be passed down only within
certain family lines.

Indeed, what we refer to as *inheritance* and what is now virtu-
ally synonymous with *genetic inheritance* used to entail a rather

loose, imprecise, and changing set of influences. Gods, weather conditions, the mother's dreams at the time of conception, and several other factors were mobilized to explain the newborn's appearance and characteristics, which today we call the *phenotype*. Reproduction was perceived in a deep sense as a complex, unforeseeable process that always involved more than the two parents alone.

From the standpoint of the history of science, the term *inheritance* was imported from the legal system into biology in the seventeenth century.[9] Today, more than fifty years after the discovery of the double helix, it appears as if the identification of the genetic basis of inheritance has separated this—necessary—component of the process from the equally necessary others. And this is one of many examples of a more general trend that is explored in this book. The social context is split off from the genetic core. Indeed, the power to shape the social context in the first place is attributed to the genes, as well.

Our hypothesis is that the more we know and learn about our own biology, the less we are able to fit this knowledge into a coherent whole. The problem does not lie in the incompleteness of our knowledge; incompleteness is a constant. Rather, in the process of the molecular reduction of our functioning as persons, the knowledge thereby gained takes on an increasingly essentialist form. The isolation and extraction of "epistemic things"[10] from their context is a necessary precondition for being able to visualize, examine, and manipulate them. What is thereby lost and made invisible is the context, including the societal context, in which they function. In this sense, we assert the somewhat heretical claim that, even if they are not scientifically tenable from today's viewpoint, earlier interpretations of reproduction were

better able to grasp the continuum of events that shaped its outcome and to connect the various factors with each other. None
of them was given unambiguous primacy. There was a continuum that provided scope for gods and wet nurses, for immaculate
conception and multiple fatherhood, whereas today the genetic
view of things prevails exclusively. The ambiguity that used to
exist proved to be astonishingly flexible in relation to the changing context, while today we see the prevalence of unambiguous-
ness and separation from context.

Our second hypothesis is that the perceived threat posed by
ART results precisely from those fragments of the living that
have now taken on essentialist features. In the case of a heterologous IVF, it is another man's essence, pinned down in sperm.
The other man, essentialized in his sperm, threatens the couple
by sneaking into the genome of the offspring. In the case of cloning, it is the genome as the pinned-down essence of the mother
that replicates itself. The threat thus arises once reproduction
loses its social context. Reproduction has been stripped of society
and scientifically geneticized. A "social bond" in the truest sense
of the term, which was clothed for centuries in social conventions and in love and power relationships, now presents itself in
its stark genetic nakedness.

That is why one of the reasons for the mistrust evoked by ART
and other biotechnologies is that they make clear what otherwise
remains vague and concealed—the distinction between what is
now perceived as essential (genetics) and everything else (which
is relegated to an undifferentiated context). ART can isolate, mo-
bilize, and transport both in time and space what is today perceived as essential. The DNA of the gametes or embryos can be
placed in new, unfamiliar contexts—in other people and at other

points in time. The increased visibility of the building blocks of life is thereby part and parcel of the reductionist approach that is central to modern sciences and their successes.

The use of ART makes explicit the shifting, transport, and manipulation of the new biological forms, substances, and entities that now take on essentialist status. If a heterologous conception once resulted from an ephemeral or forbidden encounter, this became visible only when the children grew up and began to resemble a stranger or a friend of the family. Today this plays out in full public view. The highest precept, transparency, determines the official steps in an IVF procedure. Protocols dictate which predetermined path sperm and egg cells must take into the fertility clinics. The act of conception begins with a declaration of consent from the donors and recipients. In this as in all other biomedical processes, information is indispensable. Everyone is to be informed about everything, and everything is—and must be—visible and made transparent. The geneticization of conception, made visible, corresponds to the creation of social visibility that society demands.

Closely tied to the question of visibility is that of intention. Opponents of ART usually respond skeptically to the analogies between our ancestors' experience and current heterologous conception. They argue that the fact that children are born outside of wedlock is no reason to encourage engaging in this practice using technological means. This opposition mixes two arguments that we encounter repeatedly. The first argument comes from a deep-seated resistance to accepting that we are not solely at the mercy of chance but can freely decide. The opponents of biotechnologies can accept what happens by chance. But if it is intentionally brought to pass, then they find it unacceptable.

The second argument is based on the moral authority of nature: what is natural is also good. The case of Diane Blood shows how these arguments, which are related to each other, function in practice. If a married couple learns that the husband has only a few months to live and the woman becomes pregnant, this provides the raw material for a pretty story about love lasting beyond death. If the same woman achieves the same result by means of IVF, this is considered a deterring example of acting against nature.

2 THE GENETICIZATION OF ACHIEVEMENT

THEME 2   The striving for happiness and for the improvement of human abilities is ancient. Yet in the overarching continuity of this striving, the possibility of soon taking genetic paths toward human enhancement is bringing forth new discontinuities, which become particularly salient in the field of competitive sports. More and more, in ways that are increasingly hard to disentangle, the natural (perceived as what lies within the body) is intermingling with the artificial (traditionally perceived as coming from the outside). And this entwinement of the natural and the artificial is emerging as a hallmark of the striving to enhance that characterizes life today. Using the example of doping in sports, this chapter shows how, in the name of a fictitious naturalness and an equally fictitious equality, what is being pursued is an ultimately illusory purification of (natural) life. Decisive in this, both for the challenges they pose and the opportunities they offer to this work of purification, are the last two decades' achievements in molecular genetics, including the latest insights that have led to a critical reappraisal of how genes function. Today genes are increasingly understood as epigenes—mechanisms of inheritance that do not depend exclusively on the DNA sequence. Building on the latest insights from molecular genetics, we argue that the recent developments in athletics are paradigmatic in that they reveal how rigid boundaries between the artificial/technological and the natural are no longer tenable.

THE BIOCHEMICAL PURSUIT OF HAPPINESS

Unlike assisted reproductive technologies (ART), which, with some well-known exceptions, have been integrated in society, direct biochemical or genetic human enhancement has not yet been realized. The genetic enhancement of human abilities, whether feared or desired, still lies in the future. Yet in collective awareness and certainly in bioethical debates, it is already one of the most controversial designs among a great number of controversial plans for the future. Here, too, a threshold whose crossing is a point of no return is imagined. Altering our genes, so the argument goes, will mark a radical rupture in the history of humankind. But here too the question arises: is this really a fateful discontinuity? Not entirely. And yet the idea of a change in the genetic nature of humankind displays factual ruptures with what up till now was considered possible. The discontinuities arise, once again, where visibility, mobility, and action move into the foreground.

But something else is also involved. Paul Rabinow recently cited Georges Canguilhem to point out that all technologies are artificial but that this does not mean they are unnatural in an ontological or essentialist sense. The history of the cultivation technologies used in agriculture is a continuous sequence of successful or failed attempts to use technologies suitable under the right conditions and limitations to change nature. Only where technology supports an already given potential do the desired increased harvests or unexpected desirable results appear.

Technology can thus be regarded as a mode of bringing out potentials but not essentials.[1] For Canguilhem, from a scientific standpoint it is downright absurd to always speak of a "denatured nature" (*nature dénaturée*) whenever technological means are employed. The first use of a "thing" always means its "denaturalization," which, technologically, means nothing else than varying its use.[2] Applied to the biotechnologies, this means that we are only beginning to understand how polyvalent and underdetermined organic systems are. We stand at the very beginning of exploring their limits and potentials.

Our everyday life is already molded by the vision and practice of a biotechnologically anchored enhancement of achievement. For example, a booming cosmetic industry pays homage to the ideal of perfect beauty. Plastic surgery, DNA creams, free radicals, and skin stem cells are part of the standard jargon in advertising and shopping malls. The ingestion of substances that allegedly make one wide awake, highly motivated, and concentrated— dubbed "brain doping" by the media—is rapidly spreading in academic circles. The use of medications like modafinil (called "the professor's little helper") to enhance cognitive achievement raises many questions.[3] Medications that initially were developed to treat mood swings and psychological disorders are now employed as lifestyle drugs. Fifteen percent of the American students and academics surveyed in a recently published study replied that they regularly take Ritalin to enhance their cognitive achievement. Ritalin is a molecule initially used to treat attention deficit disorder (ADD). Viagra and Prozac have long since found a fixed, iconic place in the long list of other happiness-producing molecules.

These and other happiness remedies, along with countless other anti-aging recipes, intrude on our daily life without control.

Empirical surveys show that acceptance or rejection emerges in opaque and seemingly random patterns. And this occurs in a context in which there is a lack of knowledge about both the advantages and the risks entailed with the long-term use of these substances. Nor are there clear guidelines for future use. Moreover, there is no global administration or authority charged with the task of supervising the molecular paths to "better life."

That is why it is revealing to turn our attention to an area, sport doping, in which these types of problems have been taken up systematically and publicly. Here there are bodies that are charged with achieving a global consensus on guidelines and compliance and that speak in the name of a humankind that has apparently asserted its intention to do without artificial enhancement. Hence sport doping and the technological, ethical, and political debates surrounding it are especially revealing because they provide a preview of the future scenarios of our individual as well as collective engagement with human enhancement.

FASTER, HIGHER, STRONGER: DOPING FROM WITHIN

Doping in sports, considered a preview of biotechnological (that is, biochemical and genetic) enhancement, exposes all the contradictions between our society's acknowledged and unacknowledged goals in promoting the enhancement of achievement with the available methods. What emerges first of all is the fiction of equality. We know that athletes, just like other people, are not born equal and that, in biology, *equality* is a meaningless term. We also know that doping, like genetic enhancement, could equalize some existing inequalities. But this fact disturbs us: the more that equality turns out to be a fiction, the more it seems necessary to cling to it.

The moral value of sports is proclaimed not only but especially during the Olympics, which are regarded as embodying the ideal of a fair competition among athletes. In competition, all are supposed to take individual excellence to the limits of ability equally. But the motto of the founder of the modern Olympic Games, Pierre de Coubertin ("faster, higher, stronger") also displays a completely different ambition. It spurs athletes to exceed, again and again, the seemingly fixed limits of human ability. Records exist to be broken. This driving urge to do more things, and faster and better, than others is characteristic of our time. So sports can tell us much about the potential of our geneticized future.

But why, then, is doping problematical for sports? To be consistent, shouldn't it serve instead as a model and example for a similar program aimed at improving or doping our entire lives? How can limits be transcended if the practices that make it possible to achieve this transgression are at the same time condemned? The contradiction is obvious. To be able to live with it, the ideal of *faster, higher, stronger* has to be interpreted in a way that allows the desired exceeding of limits to remain within the framework of natural human possibilities. Invoked here, too, is the image of nature raised to an absolute, essential standard. Behind it stands the image of the athlete as someone who grows beyond himself while still remaining himself. The "fiction of the natural"[4] in contemporary sports, however, is quite obviously a fiction. Diets, ingenious training plans, technological equipment including clothing: none of this is left to chance, and none of it is "natural." Science and technology are already indispensable for sport. The exploration of the molecular mechanisms that tie nutrition and training to achievement will be pushed forward just as much as innovation in technological equipment. Although all of these technologies are artificial, they unfold their effects only after they are closely interlocked with the athlete's natural efforts—from within, so to speak.[5]

Yet in apparent defiance of the conflation between the natural and the artificial that is entrenched in competitive sports, sports authorities, athletes, and their fans alike appear to find it surprisingly easy to present this highly technologized area of human activity as the manifestation of natural striving. This is possible because the fictitious boundary between the natural body and the artificially manipulated one is kept under constant surveillance and subject to an equally constant work of purification. Only thanks to this process can the fiction of the natural persist and be accepted as a *fait accompli*. This is why, for the near future, genetic doping is perceived as the last frontier and definitely the greatest challenge for the purification work of the antidoping authorities, which act in the classic scheme of cops and robbers. As soon as a new doping substance or method is available, the controlling bodies set out on a race to detect it. They expand their repertoire of tests by means of which natural athletes can prove themselves to be pure. Here again, visibility comes into play because doping is effective only as long as it remains invisible. But genetic doping undermines this situation, since if the enhancement comes from within by altering genes in the appropriate tissues, it is highly improbable that this intervention can be discovered and exposed.

The International Olympic Committee (IOC) defines *doping* as the "administration or use . . . of any exogenous substance or any physiological substance in abnormal quantities or administrations that enter the body in an abnormal way with the sole intent of increasing his/her achievement in an artificial and unfair manner." Once a doping gene is stably implanted in an athlete's genome (particularly through the genetic manipulation of an embryo earmarked to be an athlete), it ceases to be a substance (much less an exogenous substance) brought into the body in an abnormal way. This gene would function from the inside and thus would have become a component of the body in its own right. The IOC will know how to adjust its definitions to the new possibilities. But that is precisely the point: genetic doping shifts the purification work so far into the interior of the body that ultimately the boundaries that are to be watched cease functioning as boundaries.

This is not merely a technical objection that could mean the end of the cops-and-robbers game in the age of biogenetics. It points (again) to the decisive characteristic of contemporary biology. As Hans-Jörg Rheinberger remarked, the living entity, whether cell or animal, becomes simultaneously the object of investigation and the laboratory in which it is investigated.[6] For it to be examined, the living material must first be modified, and in today's biology research, questions are almost exclusively posed from the inside, treating the cell, tissue, or animal as a living laboratory that reacts to our interventions and in so doing confirms or disproves our hypotheses. This view from the inside, which at the same time is also an intervention, makes it increasingly difficult to distinguish clearly between the natural and the artificial. The artificial intervention and the manipulative step do not unfold their effect until they are completely integrated into their natural surroundings—and thereby modify it.

Thus, the fictitious maintenance of the classical dichotomy between natural and artificial in the doping controversy and its inconsistencies underscore once more the power of the action from within. For example, as is well known, training at high altitudes leads *from the inside* to an increase in the number of an athlete's red blood cells—and is allowed. But achieving the same effect with blood transfusions *from the outside* (blood doping) is forbidden. The reason gene technology is so destabilizing is that it threatens—or promises—to shift the *outside* to the *inside*. Introducing into an athlete's genome a gene that leads to an increase in the number of red blood cells would resemble the effect of training at high altitudes. It would become *de facto* a process that plays out inside and that would therefore elude the ban on blood doping from the outside.

As with ART, the contradiction coming into view seems to lie in the visibility and the attribution of agency. Genes have become visible and effective actors, agents of life. And the more visible and isolated from context they are, the more powerful they appear. This is why the idea of gene doping and genetic enhancement is disturbing. Manipulation of genes seems all the more uncanny because it appears invisible to the naked eye as well as to the equally helpless technical gaze of the doping investigator, and yet it promises to deliver, through this invisibility, the most visible of effects in the guise of enhanced achievement.

In this context, the case of Oscar Pistorius is instructive. He is the South African sprinter who, with his carbon-fiber artificial legs, drew worldwide attention to the Paralympics. He lost his struggle for the right to compete in both the Paralympics and the Olympic Games, but the symbolic effect will outlive his individual case. His claim will continue to haunt current efforts at boundary drawing by asking what happens if technological improvements continue to advance so rapidly that they leave natural athletic achievement far behind. Is it conceivable that the Paralympics will host the most exciting athletic competitions, displacing the classic disciplines? Will they offer a better and broader spectrum of achievements in which the athletes must overcome their limits with all available technological means, an achievement in which prostheses stop being auxiliary aids but become parts of the body?

Oscar Pistorius is punished for the visibility of his prostheses because he publicly displays his difference from normal competitors. But what if it were possible to use stem cells to stimulate the growth of new legs that were also faster and better than the already twenty-five-year-old legs of his competitors? If the new legs consisted of muscle instead of carbon fiber, what would the purification work look like? It is thought-provoking that a society proud of its ideal of equality sees a threat to sports precisely when technology enables a handicapped person to overtake his nonhandicapped competitors.

THE IPOD AND EPIGENETICS: RUNNING IN A VACUUM

There are already many examples of a gene-centered viewpoint in sports. Researchers in the magazine *Human Genetics* claim to have found a single gene that can give athletes lasting success under various conditions: show me your genes, and I'll tell you what sport you should choose.[7] "We don't want a competitor to run 100 kilometers a week if his genes say he would do better to run only 50 kilometers a week and to lift more weights instead," remarks the doctor supervising the genetic screening program. The most important word thereby is the innocent verb *say*. Genes increasingly "say" things, or more precisely, they are made to say things. The statements have to do with the future. They predict talents and diseases and determine what we should do and what we should not do. In this discursive context, it is hardly surprising that every genetic change is equated with a change in the internal oracle. But to what degree does today's biology justify viewing the genes as an internal oracle?

In the last two decades, the achievements of molecular genetics have led to a critical reexamination of the epistemological foundations of the concept and the function of genes. Our understanding of life on the molecular level has (re-) discovered the epigenetic components of inheritance and has thereby undermined the primacy of the genes as autonomous determinants of characteristics. Genes may well be portrayed as the egoistic self-replicators that Richard Dawkins described. But as we are learning through epigenetics, that doesn't make them into the quasi-metaphysical determinants of our bodies and behavior. The expression *epigenetics* was first used by C. H. Waddington in 1959 and has taken on slightly different meanings over the last fifty years. In its core definition, it refers to all the mechanisms of inheritance that do not depend on the DNA sequence.[8] If cells divide or organisms reproduce, their descendants inherit both the DNA and a complex set of other characteristics that determines their mature form. In this light, genes are increasingly understood and experimentally probed as epigenes. Essential to a gene are therefore not only its sequence but also the degree and pattern of its activation in the various tissues that respond at various times to the various stimuli from the environment (its *epi* portion). In this way, we can "read" how each gene is regulated in its cellular context. The regulation can take on various forms, but what is most important is that the kind of regulation is *not* driven by the genes. True, it is carried out by proteins (enzymes), and these in turn are encoded by their respective genes. But the activity of these enzymes—and therefore the regulation they impose on the genes we are considering at any given moment—are the result of cellular and organismic functions.

None of this is regarded as heretical anymore. The knowledge that the environment in the broadest sense can constantly influence genes (the environment of the individual gene is, to start with, the very cell that uses it) is completely accepted today. But this knowledge has not gained enough recognition in society at large. The reason for this lies in the way that genes were initially defined, discovered, and understood. And this goes back to one of the main arguments in this book—that the kind of visibility of the newly defined building blocks of life (the way they were recognized as such to start with) influences critically society's attitude toward them. What counts is not only *that* something is made visible but also *how* it becomes visible. Thus, the understanding of genes as codes and of DNA as the "language" of life displays a close relationship to the epistemic and societal discourse of the information age in which molecular genetics took its first steps. It is within this information paradigm that in the 1970s, for the first time, genes were manipulated inside organisms—initially bacteria and later flies, worms, mice, and eventually primates. And this manipulation enabled and sustained the view from within that we have described above and that continues to propel contemporary biology. Throughout this process, the gradual mapping of genes within the complexity of living cells and organisms went hand in hand with an understanding of them as, literally, information bearers. That went so far that today we are confronted with two different meanings of *gene*. On the one hand, genes are just discrete segments of the DNA chain that can be given a precise molecular description and that function on the population level as markers for the probability that a particular sequence is associated with a given trait (most often a disease in the biomedical instantiation of genetics). On the other hand, genes are regarded as causally privileged determinants of a phenotypical manifestation.

The lens of the information paradigm, through which the genes first became visible and understood for science and the public alike, still plays a great role in this far-reaching conflation of the two meanings. As historian of science Lenny Moss noted,

The engagement of textual metaphors with which to characterize genes as different from other biological matter, i.e., as a text-program, blueprint, codescript, books of life and so forth, has been integral to this conflationary construction. As text, and perhaps only as such, genes can be conceived as molecules and yet evade the circumstantial contingencies, the "fateful winds," which most pieces of matter find hard to resist. It is as matter-text that genes and DNA ascend to the status of sentiency and agency, as matter with its own instructions for use and, furthermore, as the user too.[9]

This conflation of meanings has wide ramifications. As long as we regard life as the result of a code script that carries itself out, editing this script must seem like a Faustian undertaking. But our digression into epigenetics and its experimental results has shown that the status of simultaneous instruction manual *and* user cannot be located within DNA as such. The script is nothing without its context, which interprets its instructions in a way that is never wholly predetermined and that responds in an extraordinarily sensitive fashion to the environment. If life is about building blocks that are, at one and the same time, users of their own instructions, then these building blocks cannot be reduced to a chain of DNA nucleotides. Rather, they arise from the internal circularity and the emergent characteristics of living cells and organisms. And this brings us back to blood doping and high-altitude training. Both regulate the same set of genes responsible for an increase in red blood cells. Only by means of a radical, if fictitious, separation between *outside* and *inside* can one be allowed and the other forbidden.

The events surrounding the New York City marathon in November 2007 illustrate this point. The responsible authorities decided to ban the use of iPods during the race on the grounds that, as a kind of emotional doping, music could improve runners' achievements and give them an unfair competitive advantage. Music's calming or enlivening effect on human behavior, in both physical and emotional terms, has been observed throughout history. Music has been used to increase human stamina from the first hide-covered drums to the iPod.[10] The decision by the marathon authorities is interesting in two ways. First, it (probably inadvertently) implies the acceptance of the idea that the distinction between outside and inside is not important after all. The music from an iPod, in interaction with genes, alters achievement in the same way as do countless other practices that have already been classified as doping. Improvements from the outside function because of the effects that they trigger and let develop from the inside. The technology is there to kindle potentials and not to turn inner features into untenable essences. The second point of interest in the marathon authorities' decision is that the ban on music players was supported by results from studies of metabolism and images of brain activity. These linked the act of listening to music with a large number of molecular reactions. Once again, the visibility emerging from neuroimaging studies made the difference here.

The marathon authorities' decision may be an extreme example of a trend we encounter again and again in the following chapters. The more our ability to enhance (or otherwise influence) human achievement increases and the more visible this ability becomes, the more resistance grows against it. This resistance creates an ideal of the highest degree of purity, an absolute standard of holistic naturalness in which people come to project an original, dreamlike state of nature. With the example of the iPod, we seem indeed to have reached a point of no return. Following the official argument that led to the iPod ban, now both genes and iPods will have to be acknowledged as "essences" to be scrutinized in the smallest molecular detail to function as essential determinants of results. But these supposed essences are indeed connected to each other, the music from the iPod influences entire sets of genes to bring about enhanced endurance or greater speed, and these sets of genes in turn contribute to shaping how the body will respond to the iPod tunes. Only a successful interaction of this kind brings about the desired result. The decisive point is that this flowing continuum between outside and inside, which the wisdom of earlier cultures was able to grasp as an ill-defined whole, can now be confirmed in molecular detail and become the object of sophisticated trials of visibility, as in the latest neuroimaging experiments that we have alluded to. And yet as, through the awareness of our intervention, nature becomes a point of arrival more than of departure, something that we make more than something we observe, we still seem to need to dream the dream of naturalness, against whose denaturization the antidoping battle represents simply the tip of the next iceberg to emerge.

It will soon no longer suffice to condemn anabolics, but genetic doping and iPod tunes will also have to be forbidden—all in the striving toward the self-purifying ideal of natural life. But the more this work of purification penetrates into the natural body, the more life becomes a puristic nightmare. What remains is a fiction of life that begins to dissolve in a mirage of absolute standards.

This development is reminiscent of and consistent with the all-pervasive and downright obsessive demand for transparency in social life. The addiction to the need to disclose everything, which drives a transparent politics toward transparent procedures in transparent institutions as a highest (but illusory) ideal, seems to be the correlate of the increased visibility and transparency of life in the molecular age. The developments in the realm of competitive sports paradigmatically exemplify the reciprocal interaction of these two influences—the scientific and the social. Banning all technological aids, including music, from a marathon competition means striving for a pure athletic body that is technologically naked and socially transparent.

3 DISPUTED AFFILIATIONS: WHO BELONGS TO WHOM?

THEME 3   What and who belong to whom? These questions of identity, property, and affiliation run through all encounters with biotechnology. They underscore the fault lines along which the public debate imagines, articulates, and decides on diverse and disputed designs for the future. The processes on the molecular level in which genes, cells, embryos, clones, and other entities are made visible and mobile lead on the societal level to appropriations that are as controversial as they are in need of clarification. Interdependencies on the molecular level are translated, on the social level, into interdependencies between individuals and the community. In this chapter, we examine two aspects of disputed affiliations and appropriations. The first revolves around disputed lines of descent that give visibility to unsuspected tensions between the individual and the community. The second aspect deals more directly with disputed appropriations, in particular with issues of property rights over genes and other biological entities. What belongs to whom also decides, in the end, who belongs to whom.

CONTESTED LINES OF DESCENT

One of the central visions in the current life sciences is the hope to decipher the genetic factors that underlie various diseases and behavioral dispositions. Genetic constellations contribute to some of the behavior patterns of individuals and in large part to their susceptibility to diseases. Although no one doubts the massive importance of the environment in the broadest sense for the shaping of human existence, the emphasis in research in the last few decades has increasingly shifted toward the individual's genetic constitution as the final cause of traits, including diseases and behaviors.

The reason for this is simple. The genetic constitution is incomparably easier to grasp in its entirety than are the diverse environmental factors that make an individual biography unique. DNA is an arrangement of four chemical units that can be expressed by the letters A, C, T, and G (as the four nucleotides are commonly referred to), which makes it technically possible to read them as text, to compare individuals, and to calculate their differences on the population level. Another significant characteristic of genes is their lifelong stability. If a person is born with the sequence AATTTC in a particular gene, it is very probable that he or she will also die with precisely this sequence (in precisely the same position). It is thus not only easy to obtain knowledge about a person's genes, but this knowledge is also easy to calculate, record, and archive.

To investigate individual variations in a population, it therefore suffices to record a person's uniqueness once. The saved sequence remains a valid signpost of a person's uniqueness even for other, future purposes. In contrast, registering all the environmentally dependent aspects of disease and behavior is simply unimaginable because to do so, the many environmental variables that influence an individual life would have to be continuously gathered, archived and computed.

Because acquiring and storing the data that make modern human genetics meaningful is fairly easy, all biomedical projects that analyze populations on the basis of their genetic profiles create a double bind and bring the person and the community into a mutually dependent relationship. On the one hand, such projects need large numbers of people who share their DNA, health history, and family history. Only this information can serve as a valid reference framework, and only the members of this community, with all likelihood, will stand to profit directly from the recorded genetic information. On the other hand, the information about the community is useful to an individual only if his or her own genetic uniqueness is set in relation to the respective population. The information that people carry in their genes thus has no meaning for them if no reference population is available in which specific DNA sequences are associated with a certain probability of the respective clinical picture. This is what creates the inevitable interdependence and the tension between each individual and the respective community that serves as reference population. As is shown in this chapter, this interdependence is also a source of tensions over ownership and appropriation.

ICELANDIC GENES: THE TENSION BETWEEN THE INDIVIDUAL AND THE COMMUNITY

DeCode Genetics Inc. is a biotechnology company in Reykjavik. Shortly after its founding in 1996, it became synonymous with the individual and collective problems raised by genetics. The idea on which DeCode's scientific ambitions and business plan were based was simple. For centuries, the Icelandic population has lived in relative isolation. There are detailed and reliable documents on the descent and origin of most families. In addition, the relative smallness of the country and its well-organized health care system made it possible to set up a register comprising the entire population. In it, three kinds of information could be stored on all of Iceland's citizens—individual data on illnesses and current state of health; centuries-old family genealogies; and genomic data (an adequately large collection of tiny DNA fragments—genetic polymorphisms—able to distinguish individuals and trace their lineages unequivocally). DeCode's Web site puts it in a nutshell: "A population with all three sets of data—genetic, medical and genealogical—is a scarce resource in human genetics."[1]

The further procedure is equally simple. People with a specific disease are registered together as a group, and their ancestors are traced. Then their genome is examined to determine what genetic variants cluster over time together with the transmission of a given disease. A much higher frequency of such variants can be expected among those who carry the disease than among the healthy control group.

From the beginning, it was clear that a project of this scope would raise new ethical questions. What measures should be taken to preserve the privacy of medical and genealogical data when creating the profile of an entire country comprising only a few extended families? What does it mean to put this information in the hands of a single actor—a newly founded biotech company that is headed by an Icelandic scientist and was financed by venture capital from the United States? Who would or should profit from this countrywide genetic exploration, which entailed the prospecting of a completely new raw material—half of history and half of the DNA of the inhabitants? For all these reasons, the Icelandic case is a prime example of two instructive aspects of contested affiliations—(1) the tension between the individual and the general population and (2) the empowerment of genetically aware citizens.

We have already pointed out that the individual and the general population are in a mutually dependent relationship, interlocked in a knowledge that becomes meaningful for the former only insofar as it refers to the latter. In addition, genetic knowledge and its applications also create new communities, and some of this community-building displays clear features of a bottom-up process. Patient and self-help groups oriented toward specific diseases, for example, are increasingly making use of the potential of the new genetics. These persons make direct contact with scientists working on their respective diseases and influence research priorities by raising funds, recruiting other patients, and taking part in sampling and distributing biological material within large multinational studies.

These communities are held together by a mixture of hope and research results, and through DNA, new relationships also emerge, and new commonalities become visible. A poignant example of this occurs when families whose children appear to suffer from the same disease discover by means of a genetic test that the children all show the identical kink on the short arm of one of their chromosomes or the same mutation in one of their genes and thus belong to the small group of people with this rare condition. When the families meet to exchange their experiences, an astonishing similarity in the children's appearance and behavior emerges. For families that have never seen each other before and are not kin, one moment is enough to forge a new relationship, as if the revelation of genetic commonality also makes visible a commonality of flesh and pain that has so far gone unnoticed.

But something else is emerging that points beyond the specific common destiny—a progressive empowerment of these newly genetically aware citizens. In 2000, the specialized journal *Nature Genetics*[2] published a scientific paper on a newly discovered disease-producing gene. For the first time, among the listed authors was the name of the activist patient on whose initiative and with whose collaboration the study had been undertaken. For a highly technical scientific journal that publishes the work only of qualified practitioners, this was a significant step. A recently published editorial article also championed the idea that, in some cases, patients should be recognized as coauthors of genetic studies.[3]

Now let us compare this bottom-up approach of empowering patients with the Icelandic case. Here, a crucial and conspicuous difference is how individuals join a genetic community. At first sight, the Icelandic case seems to have a top-down approach in which external actors assemble a genetic community. But the picture is more nuanced than that and raises complex questions. The creation of a databank comprising all of Iceland was one of the most controversial points when DeCode Genetics Inc. was founded. In 1998, the Icelandic parliament passed a law, the Health Sector Database Act, that created a national database and transferred its founding and operation to DeCode Genetics Inc. The nub of the heated controversy was that this law was based on the assumption that all citizens would automatically agree to make their data available to the database. Only after public outcry was a clause introduced that permitted exceptions. But even with this clause, it was still assumed that individuals agreed to have their data in the database unless they objected explicitly in writing. In effect, the law stipulated that the entire population of the country agreed to the data mining of its collective health record. Despite criticism of this procedure, it could be argued that the law ultimately represented the democratically achieved expression of the population's will and was thus a bottom-up procedure. But the answer to what should be understood by *bottom up* can be different, as the following example shows.

The problem that ignited resistance from the population in Iceland was solved differently in the case of the Human Genetic Diversity Project, which aimed to collect a worldwide sampling of genes to map the entire spectrum of human genetic variation. From the beginning, gaining consent to DNA sampling from native populations in remote regions, like the tropical rainforest, was seen to create special problems. The coordinators of the project therefore developed the argument that the genomes of members of ethnic groups in the rainforest belonged not to those individual members but to the entire community and that the concept of individual consent that had been developed within the context of health care ethics in the Western world could not be properly applied to indigenous tribes. Apparently the question of the ownership of genetic lineages could not be settled in the same way once and for all. What is individual and what is collective must be determined on the basis of the local meanings of these terms and of the historical and present contexts.

THE SALE OF LINEAGES: MAKING PROFITS WITH BREAST CANCER GENES

The questions we have addressed so far have been: Who may speak in the name of my genome? The parliament or me? My tribe or me? But the question of affiliation includes that of the property rights over genetic information. The Icelandic parliament sold its citizens' data to DeCode Genetics Inc. DeCode, in turn, concluded an agreement with the giant pharmaceutical company Roche to grant licenses for exclusive access to the database and thus to the jackpot of disease genes (new drug targets) that could be discovered. But what gain should the Icelandic population have from this? Should the Icelanders profit from the riches that would flow from their genes—an estimated $14 billion (U.S.)? Apparently not. The law stipulated only that the Icelandic population should have free access to the medication treatments resulting from DeCode's use of its DNA information. For many, this seemed too little and too vague. If a company can make profits from its knowledge of an entire population's genome, it was argued, then the citizens ought to receive the resulting health care benefits and also participate in the financial gains.

Some answers to the question of the ownership of genes have already been established all around the world, although in very different ways. And although a full assessment of these various settlements is beyond the scope of our analysis here, the following story highlights an important trend in how contested ownership of genetic information could be regulated in the future. The story begins with the discovery of a chromosomal segment on which a marker is found that correlates with an increased risk of developing breast cancer. While the discoverer, Mary-Claire King, and a worldwide consortium are intensively chasing the underlying gene, scientists at the University of Utah are busy founding a company to catalog the genetic family trees of Mormon families (Myriad Genetics, Inc.). In this case, there is again an agreement with a giant pharmaceutical company, this time Eli Lilly and Company, which has joined Myriad in the race for the breast cancer gene that has been preemptively named BRCA1. Myriad/Eli Lilly is to receive an exclusive license to market all diagnostic and prognostic testing procedures that result from the discovery. In 1995, both BRCA1 and its sister gene BRCA2 were completely sequenced, and Myriad/Eli Lilly began to offer its diagnostic tests.

Now comes the interesting turning point. Although both genes have been patented, the company did not choose to protect its patents via the usual path of selling licenses for the purpose of screenings to other companies or institutions.[4] Instead, it decided to produce and offer on its own the diagnostic test, thereby creating new relationships between patients, their genes, and doctors. A declared reason for this seminal decision was the goal of setting up a company-internal database of BRCA1 and BRCA2 sequences from patients who underwent the test. The company thereby hoped to be making an important investment in its own future. Establishing a proprietary genomic database of the genetic variants that correlate with breast cancer could serve as a starting point for developing new medications and even better diagnostic tests. What was originally a legal obligation or a by-product (the test provider's archiving of the test result) of a genetic test aimed at providing actual diagnosis or prognosis for a specific patient here and now was quickly becoming something altogether different—a collective asset that was both scientifically and financially open to its future potential.

In the United States, Myriad launched an energetic campaign to advertise its test to both doctors and patients. This was also an epochal shift in the history of genetic tests because previously they had been accessible only within a relatively small network of professional experts (human geneticists) and specialized institutions (national referral centers). This critical change enabled, in turn, a thorough reconfiguration of genetic lineage relationships (in this case, those having to do with breast cancer), but that speaks to more general features of today's encounters with genetic knowledge. This is because the decision of Myriad/Eli Lilly to market the test directly and make it available to a broad public meant that people without a family history of breast cancer could now also be tested for BRCA1. In this way, assuming that they decide to submit to an examination to prevent the uncertainties of an open future, people could enter through their DNA into a new kind of collective, Myriad's BRCA1 archive, regardless of whether they had a clinical indication for undertaking the test and independent of its result. And so we are back to visibility: the previously invisible genetic material of individuals turns into a visible, collective—and potentially profitable—genetic resource. It is visible, collective, and potentially profitable as an indispensable reference framework for future predictions that express in percentages the risk of falling ill.

And so we can now retell this story from a different angle—that of one of the first women in whose blood Mary-Claire King identified the region of chromosome 17 that harbors the gene for increased predisposition to breast cancer. Thanks to the genetic material that a large number of people had made available in public databases, it was possible to home in on the right gene. Dense interactions between science and the market and between publicly and privately funded research are a hallmark of many biotechnical innovations. In the same moment that it received its molecular identity, the completely sequenced gene immediately became the object and property of a single company. When the BRCA1 gene emerged from the messy mixture of chromosomal substances buried in individual bodies as a visible text with mutations that could be read, recorded, and patented, it began writing its own story. Once again, a publicly accessible object of scientific inquiry was transformed into a private object, the bearer of exclusive and therefore contested property rights.

A DINNER WITH CONSEQUENCES: HOW TO GOOGLE YOUR GENES

"Eventually, there may be true insight into the relationships between nature and nurture, and the individual will then benefit from the contributions of the community as a whole."[5] This is the final sentence of a trailblazing scientific article that described for the first time the complete sequence of an entire human genome, that of Craig Venter, the entrepreneur and scientist who in 2001 started to compete with the International Human Genome Project in the race for the first draft sequence of the human genome. Against all expectations, the race ended in a draw; and when President Bill Clinton announced the decoding of the human genome, he was flanked on one side by Francis Collins, the director of the International Human Genome Project, and on the other by Craig Venter, the chief executive officer of Celera Genomics. The picture of these three men is a symbol of the place that the life sciences have acquired in the public realm. It expresses their political significance as well as their convergence with the market economy and the forces driving it.

Until recently, however, the full sequence of the human genome consisted of a composite arrangement of the genomes of different people—one piece of chromosome 1 from person X, one piece of chromosome 2 from person Y, and so on. It was the sketch of an idealized human genome formed from the assembly of individual differences. But if the challenge of comparing individual genomes is to be taken seriously with the aim of setting their variations in relation to various diseases and traits, then a quantum leap forward is needed. Borrowing from the textual metaphors analyzed in the previous chapter, the point is now to read what complete individual genomes actually say. The first such attempt already led to unexpected results. Venter's genome differed greatly from the genome that until then had served as a reference. Most of the differences were not simple changes in individual chemical letters but changes in complex rearrangements of whole groups of letters (what is called *structural genetic variation*).

This discovery is important because it takes us back to the visibility of life's molecular units. Previously, most studies of individual genetic differences assumed that the most frequent and most important changes were single substitutions in the individual units that make up DNA (for example, a C in place of a T, and so forth). Those were the changes that were looked at, and not surprisingly, those were the changes on which researchers derived estimates of the degree of genetic variation among humans. Now Venter's DNA, the entire sequence of a complete human genome, forces us to rethink this dogma and to focus on types of variation (the complex rearrangements mentioned above) that have escaped our attention simply because we have not been looking for them. Today it has been concluded that the assertion that only a tiny part of people's DNA is responsible for genetic differences was imprecise.

But other questions arise that go beyond the scientific and technical details of Venter's genomic sequence. First, this quantum leap in our understanding of human genetic variation was financed solely by the private sector—indeed, by a single person who became a scientist or by a scientist who became an entrepreneur. But Venter made this private achievement available to the public domain as a resource. This includes the symbolic gesture of choosing to publish the results in the specialized journal that stands for open access—*PLoS Biology*. The aggressive patenting of random genes (which Venter once advocated) may have had its day after careful deliberations about what is at stake if the dissemination of important knowledge is limited. This new strategy apparently builds once again on securing and legitimating the public interest in scientific progress. Beyond the public display of private data (and private money), a completely new aspect has emerged in the tensions between public and private and between the individual and the collective dimensions of modern genetics.

A private dinner in February 2005 at which Craig Venter, Sergey Brin (the cofounder and president of technology at Google), and Ryan Phelan (the CEO of DNA Direct) took part provides another insight into possible future developments. As one of the participants reported, it was a dinner focused on "googling your genes."[6] Google's unique asset and the basis for the company's near monopoly in Internet navigation is its incomparable computer capacity, which results from joining a large number of personal computers in a single huge computer that is able to carry out an almost endless number of search operations in a very short time. A few years ago, Google agreed to support a megaproject at Stanford University whose goal was to simulate *in silico* the process of protein folding *in vivo*, one of the most complex biological phenomena and an extreme computational challenge. Google provided the greatest part of the computing capacity and allowed its worldwide clients to use Google's search toolbar to register with the Stanford Project. While the users slept or were away from their computers, the latter continued working for Stanford: the gigantic computer organism breathed in the countless laptops of its voluntary helpers.

This experience immediately made clear to Venter that Google possessed the computer capacity and the vision for its constant expansion that were needed to master the imminent flood of genetic information. Once complete sequences of entire individual genomes are created, the crucial challenge lies in the ability to actually manage and analyze this previously unconceivable wealth of data. Sergey Brin was willing to accept the challenge and test Google's capacity to search and calculate. In Venter's words,

People will be able to log on to a Google site using search capacities and have the ability to understand things about themselves as they change in real time. What does it mean to have this variation in genes? What else is known? And instead of having a few elitist scientists doing this and dictating to the world what it means, with Google it would be creating several million scientists.[7]

In this framework, it was logical that the CEO of DNA Direct, an Internet company that offers genetic tests and consultation, would also be invited to dinner. DNA Direct recruits its customers/patients with advertising that it purchases from Google using a simple schema. When a person searches in Google for a specific disease, a window opens on the upper right corner of the screen and offers the genetic test for this illness.

So here is a confluence of science, information technology, and private capital that effortlessly transcends all surveyed boundaries between private and public, individual and collective. In the vision of "googling your genes," the personalized medicine that promises methods of treatment customized to the patient's genetic profile appears initially and primarily as an exercise in the democratization of science, since it enables people all over the world to gather information about themselves and to use this information accordingly. But what does it really mean to learn things about oneself and about how those things change in real time? To what extent does this influence our perception of the self and its relationship to the community?

Beyond the scientific reception, the psychological influence of Venter's genome and its symbolic power is rich in consequences. The article in *PLoS Biology* presents on a single page a person's complete genome, compact and potentially readable (though only with an especially good magnifying glass). For example, one can learn that Venter has a predisposition for fluid earwax (a small detail) and also has a disposition for cardiopulmonary diseases and alcoholism (more important traits). These are precisely the kinds of details that give the project its ambition and its fascination. As consoling as it is uncanny, as promising as it is disturbing, DNA is creeping into the routines of everyday life.

At the end of the previous chapter, we trace how genes came to be understood as text. The most-far reaching consequence of this has been the conflation of two completely different meanings of *genes*—on the one hand, probabilistic signposts of future fates and on the other, causally privileged, essencelike determinants of biological outcomes. The circle of that conflation now closes when one's own genetic text becomes searchable and comparable through Google. When Venter explains that "an impressive collection of large sets of genes, together with environmental conditions, is what determines our life," we are still left with the difficulty of capturing this image. And the presentation in *PLoS Biology* does not help us much. The linear map of the genome invites us to browse it as our own book of life, and yet it is ill-suited to capturing the interwoven, probabilistic network that we are being asked to make sense of. In fact, this linear map—this two-dimensional cartography—is compatible with the Zeitgeist of Google. And yet it is currently unclear whether the simple linear representation of the genome, as conveyed in public and reinforced by the use of Google, can be replaced by a more integrative viewpoint of the interplay between genes and environment.

The decisive challenge is to make understandable to the public that scientific knowledge can change rapidly and that the latest state of knowledge is always a preliminary one.

In conclusion, the anarchic vision of a great number of scattered individuals searching the Web for the meaning of their genetic uniqueness—which will be possible in ten years for about $1,000—is very superficial. Neither the "elite scientists" who offer exclusive knowledge about the genes (along with paternalistic advice) nor the "millions of scientists" who are envisioned suddenly emerging from googling their genes seem suited to realize the existing potential knowledge in a societally sustainable way. New political approaches are needed to channel these changes—because individual genomes become the source of individual destinies, because individual citizens can give this knowledge meaning only if they associate their own uniqueness with the collective aggregate as the sole statistically relevant source, and because private interests thereby intersect intimately with pubic interests.

4 CONTESTED FUTURES: EVERYDAY EXPERIENCE AND THE VALUES DISCOURSE

THEME 4   The question arising for society is how the forms of life created with the molecular approaches we describe in this book are to be integrated into the existing social order. From the wealth of scientific-technological potential, what can and should be realized? Three discourses provide partial and contradictory answers to this question—innovation discourse, risk discourse, and value discourse. The latter, in particular, is molded by a flood of uncontrollable images that create associations of the new forms of life, shaping people's imagination and everyday experience. Many of these images are changing a centuries-old perception of nature as an unchangeable source of moral authority. Nature is reduced to matter that can be manipulated and patented. But this viewpoint provokes counterforces that want to cling to the image of an unmodifiable nature. In this fissured moral landscape, the genetic visibility of life encounters the visibility of (reenergized) values. But in a pluralistic society, values are and remain heterogeneous. They change and stand in contradiction to each other. The struggle to shape the future has begun, but however open and uncertain it may be, it must be conducted in a way that makes living together in a pluralistic society possible.

FROM PRESCIENTIFIC EXPERIENCE TO TECHNOSCIENTIFIC
DAILY LIFE

According to Gaston Bachelard, the guiding idea that modern natural sciences pursued during their institutionalization in the seventeenth century was a progressive distancing from everyday experience.[1] Since then, people's everyday experiences have been sharply and irrevocably separated from scientific experiences. The progress of science proceeds against everyday knowledge, which contains nothing but epistemological obstacles, and the emancipation of science from everyday experience thus occurs by overcoming errors—not by presupposing the given but by creating its objects and construing its own experimental systems. This presupposes the deconstruction of objects of everyday experience—a radical epistemological rupture (a *coupure épistémologique*). The progress of scientific development is thus characterized by the separation of prescientific everyday experience from scientific experience. This rupture leads to thinking in terms of radical discontinuities, although Bachelard acknowledges the creative forces of the imagination and of images. There appears to be no more need or room for the everyday observations and the many minor technological interventions with which people have interpreted and altered their environment for centuries. Bachelard underscores that this rupture from everyday experience was at the same time the precondition for the autonomy of science and its protection from the grasp of political and religious powers.[2]

In the preceding chapters, we repeatedly refer to the continuities that often are buried or pushed aside and that disprove the notion that science creates only radical novelties. The new often just happens to veil its origin and antecedents. To render it possible to adapt to the new, similarities with the old have to be produced. Discontinuities are thus always constructed also to highlight especially relevant and sensitive characteristics. To-day science and technology, with their ingenious technologies to make things visible, shape people's everyday experience. Yet de-spite these developments, are the obstacles to knowledge thereby finally removed, or have they merely shifted location?

Although cognitive and emotional gaps have arisen between scientific knowledge and everyday experience and scientific ex-perts differ from laymen in their language, concepts, methods, and worldview, these are not necessarily irreversible ruptures. Never before in history have access to and level of education been as high and widespread as they are in industrial societies today. This opens up entirely new engagements with knowledge. The distinction between producers and users of knowledge has become blurred in many areas or is removed by numerous inno-vative steps that create new connections between the production of knowledge and the diverse contexts of its application. The political legitimacy of technological decisions depends on a bal-ance between democratic processes and the best possible techni-cal expertise, a balance that has to be found anew again and again. As empirical works in science studies show, technical expertise displays a broad spectrum of different forms. Expertise has to do with experience. Experts have to have appropriate knowledge and experience, and the two can differ greatly. Expertise includes an ability to take part in a discussion and an ability to make highly qualified scientific contributions. It can consist also in metaexper-tise—the ability to make critical distinctions without specialized knowledge.[3] It is thus no longer possible to draw a simple and lasting boundary between experts and laymen.

In recent decades, the cognitive and social authority of sci-ence has been exposed to an onslaught of societal demands for increased opening and accountability. Today, on a broad front, emerging problems challenge science and democracy to reorder their relationship. In this reordering—often in a manner that may appear self-contradictory—three discourses mold our ideas about shaping the future. The first is the innovation discourse. Here the aim is to shape the production of knowledge in a way that, under suitable institutional and financial conditions, science produces knowledge that sooner or later will be technologically implemented and economically profitable. Under the unceasing pressure of global economic competition, contractual agreements are laid down in milestones and deliverables. Their fulfillment is assessed with a variety of monitoring, evaluation, and auditing procedures. In an ever-denser net of regulations, transparency is raised to the status of primary criterion. In turn, this is expected to restrict within calculable and politically tenable limits the free scope that is a prerequisite for the unleashing of scientific creativity. The innovation discourse fosters all new knowledge that promises economic growth and well-being. Everyday experience mostly keeps up with technological developments and, from the perspective of the users, gives them innovative impetus.

The second discourse is the risk discourse. It has had a crucial part in shaping the changes in the relationship between science and society in the recent past. Everyone affected wants to be included in a democratically legitimated way in decisions about technological and scientific developments that entail real or suspected risks and in appropriate communication and evaluation of risks. Earlier controversies focused on differing ideas about which forms of society are worth living in and striving for. Today, a shift has begun from ends to means, as in other governance arrangements. To be able to master conflicts about incompatible goals, these conflicts are increasingly silently assumed. The discussion can then be limited to means that are neutral or simpler to administer. This increasing turn to procedures is referred to as *proceduralization* and was made possible by the formation of defined principles that have entered into the established routine of dealing with conflicts, including the seemingly intractable ones that animate bioethical debates. Among the most important principles employed in this move to proceduralization are the precautionary principle and the attempt to shape new developments in a way that ensures at least a minimum of reversibility.

In contrast to a still widespread assumption, the risk discourse is only partially technical. It calls for stakeholders to be democratically included in decisions on the actual performance of regulatory bodies in the task of controlling controversial technologies and of securing citizens' freedom of choice. The risk discourse grows out of the increase in knowledge and its possible consequences. It feeds on the uncertainty inherent in the growth of knowledge. And for this very reason, it calls for intervention by citizens, whether to act or to refrain from acting. In risk discourse until recently, everyday experience and scientific-technological experience have come closer together. Their compatibility must be striven for in the future as well, in the sense of a thriving and creative societal accommodation of the artifacts created by technology.

The third discourse is the value discourse, which we delve into in greater depth. What Bachelard called *prescientific experience* often collides frontally with the scientific viewpoint. The limitations that the value discourse calls for are not based on economic utility and feasibility or on a careful weighing of options with differing risks. They have to do with fundamental decisions and the determination of principles that operate beyond a utilitarian and democratic political viewpoint. As Charles Taylor said, they are about maintaining the differences between, on the one hand,

the habituation to our everyday life (which touches the nerve of
modernity in the interplay between production and reproduction and manifests itself in naturalism and utilitarianism) and on
the other hand, something higher—a pursuit of life based on our
most important moral distinctions.[4]

Taken together, the three discourses are an impressive confir-
mation of the success of the technosciences but at the same time
are also conclusive evidence that they cannot by themselves control the circulation and application of their products. Today, society is veritably flooded with new products, technologies, instruments, and images whose existence is all a direct or indirect result
of the technosciences. They suffuse our daily life in both work
and recreation, alter perception and experience, and in so doing
create new images of our selves and the worlds that these selves
inhabit. In the face of an economic dynamic in which every university is anxious to found startup companies and acquire patent
rights, science has long since lost its monopoly on interpretation
and definition. And in the controversies over technological risks,
science also lost public trust because it was unable to preserve its
credibility in the political arena. Thus, prescientific experience
and everyday viewpoints are no longer an obstacle to the development of scientific theories, as Bachelard believed, but people's
everyday worlds are flooded by a "second nature" created by the
technosciences. Today, prescientific everyday experience is satu-
rated with science and technology, and everyday life has become
a scientific-technological everyday life. But that does not yet answer the question of the foundations for shaping the future.

Craig Venter's dream of the instantaneous arising of "millions
of scientists"[5] who buy into the offer of a worldwide corporation
to learn more about their genes may be farfetched and certainly
underestimates the importance of expert knowledge and its interpretation. Yet the articulation of that vision makes clear that

much of what circulates in the realm of everyday experience, even when originally a product of technoscientific ingenuity, easily eludes scientific quality control. Scientific-technological everyday experience therefore requires some form of scientific accompaniment or chaperoning. To ensure it, science must go to the places where its products are used. It has to move into the everyday life, the working world, and the health provisions of millions of people who have hardly any prospect of ever becoming scientists. Seen from this angle, prescientific everyday experience, from which science initially separated, has caught up with science again as a result of science's achievements in everyday life. The epistemic rupture itself seems to be dissolving.

THE FLOODING OF THE PUBLIC SPACE WITH OBJECTS AND IMAGES

Biological entities and objects taken out of their context still follow the relatively tractable (if also confusing) trajectories that the state regulates and the market increasingly dominates, but this is not true of the images that render them visible to people and often constitute their public existence. Whatever images do—how they perform or what they represent or depict—they decisively shape the presence of the life sciences in society. They do this in what is called objective reality as much as in the imagination. The three-dimensional ultrasound scan discussed in chapter 1 is advertised and sold. But its performance by far exceeds its technological promises. By stimulating the imagination of parents and media consumers, it eludes the firm grasp of the scientists and engineers who invented it.

That is how images and concepts associated with them flood the public realm. The moment that the concepts and the phenomena they refer to have left the laboratory with its specialized terminology or escaped the discourse of the experts in ethics commissions, they begin to take on an uncontrolled life of their own. Most people associate them with other images and words they are already familiar with. The presence of these images in the public media turns the unborn child into a "latent citizen" (according to Sigrid Weigel) and reveals the ephemerality of the epistemic rupture. As Sarah Franklin notes:

> The anxious attention so often directed to "the" embryo, as in the perennial debate over "the moral status of the human embryo," forgets that human embryos are now a vast and diverse population, imaged, imagined and archived in media as diverse as liquid nitrogen, DVDs, virtual libraries, t-shirts, logos and brandnames.[6]

If no one—state, religion, science, or even the media—possesses the hegemonic power that determines what the images represent and how they are to be interpreted, then the dam is broken that once separated the experience of everyday life from knowledge about what Lewis Wolpert calls the "unnatural nature of science." The second nature generated by the technosciences asserts its independence. The images that owe their existence, their ability to circulate, and their aesthetics to the visualization technologies of science have emancipated themselves from their creators. Before we show how the flood of scientific objects and their images collide with a counterflood fueled by values, we must first register the changes affecting the most important image of all—the image that humans make of nature.

NATURE HAS NO SECRETS ANYMORE

In his book about the three most famous words that have come down to us from Heraclites—*physis kryptesthai philei* (nature loves to conceal itself)—Pierre Hadot traces the fascination exerted for centuries by the idea of a nature that hides its secrets from human view.[7] Disrobing nature, which is often iconographically depicted as a woman or goddess, to wrest her secrets from her or to outwit her is a recurring theme and a primary motivation for (male) researchers, as described by many of them from Francis Bacon in the seventeenth century to François Jacob in the twentieth century. After his historical review, Hadot notes that nature has no secrets anymore. All that is left is for philosophers to contemplate the riddles of human existence.

It is not that there is nothing left to discover or that the desire to reveal secrets has been extinguished. The limits of science are still far from view. But together with the iconography of the goddess who was to be unveiled, the idea that she is enigmatic has also vanished. Nature seems to have nothing more to hide from us as spectroscopes, sequencing machines, centrifuges, very large telescopes (VTLs), and the large hadron collider (LHC)—all of them assisted by enormous computational capacity—make the smallest and largest in the universe accessible to the human gaze. For the life sciences, knowing life means changing life (and changing life means knowing it anew). Synthetic biology even claims to be able to reassemble life in new ways from standardized building blocks.

All of this technological prowess affects our image and ideas of nature. This image was always an ambiguous one. Nature was cruel and lovely at the same time. It protected people, who simultaneously had to protect themselves from it. It threatened them and therefore had to be subjugated. Today, however, the dominant idea is that we have to protect nature from people—for example, by conserving biodiversity such as rare animals, rare plants, and threatened languages. The grand challenges of climate change, energy, the environment, and the struggle for scarce resources underscore the economic side of this often nostalgic image of nature. But above all, what the life sciences do with nature is made obvious to us every day. They intervene and manipulate, they disjoint living organisms into the smallest possible units and implant the latter into other organisms, and they breed in petri dishes forms of life that have never existed before. When, after many failed attempts, Ian Wilmut succeeded in breeding Dolly the sheep, the decisive aspect of the "Dolly technology" consisted in dedifferentiating the function of the cell. Not quite ten years later, Japanese scientists managed for the first time to reprogram skin cells into an embryo-like state by manipulating just four genes. We continue moving further in a direction in which even biology's temporality can be outwitted.[8] Another step in the direction of reversibility of life has been taken.

So what remains of the centuries-old image of nature, which gave people a purchase and orientation, a naturally given order (created by God or the gods) whose laws stood above human law and the political order and was therefore absolutely valid? Thanks to progress in the life sciences, processes of reproduction and nutrition that were regarded as immutable have changed and even been made reversible, the clear separation between life and death has blurred, and the phantasms of immortality and of endless human improvement have thrived. What remains of nature as an unchangeable authority to which humans and their laws must submit?

The decisive point is that nature has always also been regarded as a moral authority. For millennia, whole civilizations imposed on human behavior and communal living arrangements the dictates of a natural order as an imperative, accompanied by the symbols and rituals of the various religions. All this has collapsed in the last few decades in the face of scientific-technological achievements. Nature has ceased being a moral order in the double sense of the term: it is no longer regarded as a world in which eternal, unchangeable principles rule, and it is therefore less and less in a position to impose these principles on humans.

THE RISE OF VALUES AND THE RETURN OF RELIGION

When objects and images flood the public realm, become associated with those already inhabiting the imagination, and escape from the control of science, the moment of invoking values has arrived. In the face of the collapse of the idea of nature as the valid order, the value discourse mobilizes symbolic resources or attempts to resurrect some that have been buried. On the surface, the focus is on bioethical issues arising from the unprecedented scientific-technological ability to manipulate, intervene in, and control life processes.

But the value discourse by far exceeds bioethical questions. It is about conceiving the future from a perspective other than the one offered by the innovation and risk discourses. But where is the decisive difference? A first difference is that something unheard of is postulated for living together in society—the existence of nonnegotiable values. At its extreme, the value discourse insists on principles that claim absolute validity. It does so on the grounds of reclaiming the precedence of goals over the means to attain them. The question of how we want to live replaces the question of how we can get to this state. This may initially seem a good thing. But in reality, this not only challenges liberal democracies and their esteem for tolerance and democratically lived pluralism. A value is also challenged that has until now been widely acknowledged and cherished in liberal democracies—the freedom of scientific inquiry and research methods.

The value discourse polarizes precisely because it insists on principled decisions. Thanks to the freedom of scientific inquiry, a wealth of possibilities and options exists today among which we must choose anew, again and again. Here too it is true that every free decision includes the possibility of choosing between the objectives available. And every freedom of choice must include the means to actually enact that freedom because freedom is a value but also a resource that empowers action. The potential of the life sciences is immensely explosive because it offers options that challenge society's ability to make decisions in this comprehensive sense. The potential of the life sciences is not a projection onto the future; every potential exists solely in the present. From among the plethora of possibilities available today, only some will be realized. Therefore, realizing the potential depends on the limitations placed on it and on the stability that is thus achieved. In the nature of an open temporal horizon toward the future, Manuel Castells's "edge of the everlasting" will never be reached. We as a society need to develop the ability to use these possibilities to shape the potential, including its limitations.

If we compare the tools available to the various discourses, important differences emerge. Since the value discourse invokes given goals and nonnegotiable positions, it limits a priori the potential of the life sciences to set objectives and to apply certain means. The innovation discourse and the risk discourse, by contrast, use their rudimentary tools to explore and thereby to experiment, although in different ways. As discourses anchored deep in an ambiguous and incoherent reality, they lack a clear idea of knowing goals and means. But perhaps a fragmented, imprecise viewpoint that does justice to the open development of society is a protection against heated struggles over final positions about which today's societies cannot arrive at a consensus. Perhaps they

are therefore the better guarantors of the liberty that every potential requires if it is to unfold.

The value discourse ensued relatively late, simultaneously with
one of the most significant societal changes of recent years—the
return of religion. It was long assumed that modern societies have
committed themselves to secularization, but it now turns out that
only a few Western societies (Europe, and even there in varying
ways) serve as a model for this assumption. The strengthened value discourse comes at the same time as the increased importance
of religion in public life, which conceals the fact that the return
to religious values itself is subject to a process of pluralization.
This makes it improbable that religion can find its way back to its
former molding role in public life. As Hervieu-Léger aptly put it,
setting standards (with the contradictions entailed in them) in the
bioethical discourse turns out to be an eminently political process
that seriously undermines the cultural plausibility of the code of
meaning that religions can offer.[9]

As has been shown, the production of new biological entities, the removal of objects from their contexts, and the power
of the images inseparably associated with them are an invitation
for people to try to control them. Since scientific terminology
does not extend far beyond the laboratory, biological objects and
facts receive their names through external cooperation with the
media. These objects and facts are categorized, assessed, accepted,
or rejected—a development that increasingly mobilizes religious
feelings. Theologians and ethicists discuss what science cannot
define—when life begins and ends, what an embryo is, and how
people are to die naturally when they are surrounded by the
medical possibilities of artificially extending life. When religions
grapple with biomedicine, a kind of geneticization of belief arises. Even where faith resists science, it cannot escape it entirely.

The value discourse is characterized by the fact that the public realm is populated by a wide diversity of stakeholders, and as their number grows, so does the diversity of standpoints and the ambition to find a consensus about living together. When U.S. President George W. Bush insisted on banning the use of federal funds for research on embryonic stem cells, initiatives in California and elsewhere were formed to promote it with generous private funding. As we show more extensively in the case of the ethics council instituted by the former president, a new direction of research arose whose goal was to produce "morally neutral" embryonic stem cells to avoid hurting people's moral or religious feelings. Here pluralism unfolds its full effects when the state, for moral or religious reasons, refuses to support something that ends up prevailing in other ways and in other places. Along with California, there are also the Singapores of this globalized world where research centers strive for world-class biology. Science proves that it is possible to get around moral doubts. What it does is equally remarkable: instead of separating values from facts, as its rhetoric prescribes, it incorporates values by redefining them as social facts.

DECOLLECTIVIZATION, INDIVIDUALIZATION, AND NEW REALMS OF COLLECTIVE EXPERIENCE

The path to pluralism in the public realm was prepared by two interconnected developments. One is the seemingly inexorable rise of the individual. Historically, the great political and economic transformations of the nineteenth century led to the separation of the individual from the mass. The individual was constituted and legally recognized, not least by the statistical surveys that were an important administrative instrument for the rising nation-states. Reducing people to classifiable traits that could be normed and standardized to make people more graspable for state control paradoxically created a free space in which the modern individual could take shape. Recently, the process of individualization has spread through many realms—from work to love, from politics to art, from forms of entrepreneurial organization to altered family relationships. This trend is consolidated by the value that the individual takes on in economics and politics as a universally wooed consumer and voter.

The second development is driven by the successive withdrawal of the state from its role as primary actor in organizing and enforcing interests of the society as a whole. This creeping decollectivization stands in striking contrast to the great societal projects of modernity that inspired the nineteenth century, including its wars and colonialism. The twentieth century was shaken to the foundations by previously unimaginable catastrophes whose ideological and organizational roots lay in the great societal projects. The present has sworn off these great utopias and contents itself with mini-utopias. They are reserved for the individual and his or her right to live out personal fantasies.

A public realm decollectivized in this way was thus left to the self-determining, self-actualizing, and self-emancipating self. But even when initially thrown back on his or her own resources, the individual remains a social being, a person for whom being alone when dealing with problems is intolerable. This is why initiatives arise that form the core of a rediscovered civil society. These are the patients' and self-help groups that, in various ways, allow people to share with others their individual experiences with disease, suffering, and preventive action. Information and communications technologies actively contribute to the rise of such virtual communities of "somatic citizens,"[10] in which people shape their knowledge environment using the offerings of Internet medicine and other communications media. These and many other current strivings to invent or find new forms of living together make it clear that people need to listen and talk to each other. They depend on communicative exchange and mutual consultation. Freedom of choice is not exhausted in the individual act of deciding or in the fact that politics and economics impose this freedom of choice on individuals. There has to be something in between that makes it possible to repopulate the decollectivized public realm with joint deliberations and meaning. That is the only way that individually acquired experience can be transformed into common experience.

The public realm, from which collective experience and coping have largely disappeared (the welfare state has meanwhile mutated into a public-private venture), is becoming increasingly empty, making way for a new, still unordered, even chaotic pluralism. Today, countless people experiment with their experiences and life projects, including how they deal with illness, age, and uncertainty in the face of an open future. While the state deregulates, economically potent companies stream into the resulting vacuum with their products and offers of advice. In this seeming chaos with its tough competition for body, soul, and mind, people see a return to values as a means of gaining orientation and stability.

This, too, is nothing new. John Dewey recognized as early as 1927 that technological societies constantly generate "issues" (problems) that are resistant to solutions within the existing institutions.[11] He therefore suggested an experimental approach that explicitly permitted extrainstitutional practices. Dewey recognized that a public forms as soon as those affected by a problem organize around it and invent or find appropriate possibilities for action. That the public constituted in this way often seems incoherent and without plan is not surprising, since the existing political structures do not fit the new situation, for which no solution is yet in sight. For a public to form, it has to proceed through issue formation. Already before Dewey, Walter Lippmann argued in *The Phantom Public*:

[I]t is in controversies of this kind, the hardest controversies to dis-entangle, that the public is called in to judge. Where the facts are most obscure, where precedents are lacking, where novelty and confusion pervade everything, the public in all its unfitness is compelled to make its most important decisions. The hardest problems are problems which institutions cannot handle. They are the public's problems.[12]

This holds as well (or perhaps even more) for the controversies that dominate the value discourse. As the pragmatists correctly recognized, democracy is more of a practical necessity than an ideal. It is the smaller evil. Put differently: "The way to protect democracy is not to cheer it, which we do too much, but to reform it, which we do too little. But how?"[13]

5 SHAPING THE FUTURE: HUMAN TECHNOLOGIES OF STANDARDIZATION

THEME 5   In this chapter, we examine the complementary and mutually supporting human technologies whose task is to stabilize the social order, which the offerings of the life sciences threaten to put out of joint. At issue is how to integrate the newly created life forms and other biological entities into society in a way that permits an acceptable coexistence between humans and the artifacts they create. Human technologies have suitable mechanisms and procedures for this task, but they have to be adjusted for context. In the framework of living together in society, human technologies function like material technologies. After they are established, they replace the need to find a new consensus over and over again. They create standards that serve as reference points for otherwise incomparable situations and for the use of various means and appropriate behavior. The oldest and most time-tested human technology is the law, which, facing the introduction of new technologies, is constantly confronted with great challenges. Since the industrial revolution, society has striven to adapt technology to people's needs, rather than vice versa: the aim is to humanize the new technologies. Under the influence of globalization, another human technology has spread. The relatively young instrument of governance translates the widest possible spectrum of often mutually exclusive interests of a growing number of actors into political options and manages the resulting interdependencies. Bioethics has established itself as a third, indispensable human technology. It has meanwhile become a highly professionalized, universally valid currency in a global moral economy. Liberal democratic societies apply these human technologies with varying success when they seek to shape their contested future in a pluralistic way.

THE HUMAN TECHNOLOGY OF LAW

"There are things one gives as presents, things one sells, and things one must neither sell nor give away, but preserve to pass them on." For the French anthropologist Maurice Godelier, this is one reason why it is important, in market-dominated societies, to ask which things (and social relationships) ought to remain outside the market.[1] Neither the idea nor the principle is new, but "the things" that are now pressing into the market have changed. Law needs to define them—embryos, chimeras and other hybrid entities, genes and gene sequences, and the processes by which "things" change or are newly invented and produced—and to define their place in society. Biological entities must be transformed into juridical entities and domesticated, and property rights to them must be permitted or refused. Social relationships, including kinds that did not previously exist, must be regulated. Biology, with its reductionist approaches, breaks the connection between the whole and its parts, but law must restore the relationship between the parts and the whole. In short, as Alain Supiot has noted, the law is a technology among others but not a technology like the others.

The law has not been inactive in the past. It long ago complemented or replaced marital fatherhood with biological fatherhood and made nonmarital children the equals of marital ones. The biologically necessary mediation through surrogate motherhood of *in vitro* fertilization, which has become routine, and the rights and duties of all mothers and fathers involved in artificial reproductive technologies have been regulated. The law has recognized a man's right to become a biological father after his death. But it has also promulgated prohibitions. With the exception of only a few countries, it has not indulged the wishes of parents who want to have a child solely because its genetic compatibility with a diseased sibling could preserve the life of the latter by means of an organ transplant.

But the challenges of the future exceed what has been achieved so far. Remaining with reproductive medicine for the moment, as fewer children are born in highly industrialized societies, the expectation grows that medicine will make it possible to produce only healthy children or even to equip them with specific characteristics. The concern for the welfare of future generations that arose with the environmental movement encounters in biomedicine the unhappy heritage of eugenics. Targeted genetic influence beyond a single generation demands careful weighing, for the good of the child. The case of deaf and mute parents who wanted a deaf child underscores this dramatically. Here a court had to decide whether the welfare of the child would be better served if it is born as a member of a biologically and culturally defined community (the deaf and mute) or as a child whom society defines as healthy.

But difficult decisions are required in other areas as well. In an age when venture capital is invested in companies even before new discoveries are made or new processes are invented, the patentability of genes and other biological entities is as economically attractive as it is vehemently debated. In the legal dispute *Diamond v. Chakrabarty*, the U.S. Supreme Court declared in 1980 that living organisms can be patented and granted a patent for genetically modified bacteria. Worldwide, there are already more than half a million patents for (human and nonhuman) genes. In his novel *Next*, Michael Crichton stages a fictional pursuit triggered when a person is the bearer of a gene owned by a company and decides to go underground.[2] A gang of criminals is hired to kidnap the gene-bearer's daughter and grandson, since they too carry the gene. The story ends in reconciliation, insofar as an enlightened U.S. judge suspends the right to patent genes on well-argued grounds that are documented in detail in the appendix, whereby it turns out that the story's starting point is a true event. The patentability of genes remains controversial in the United States (in Europe it does not exist) and is currently being subjected to legal revision.

In the public realm, the patenting of genes is noted with worry. The idea that someone else could acquire ownership of one's own genes or those of others triggers widespread unease. In chapter 3, there is a discussion of a diagnostic test developed in the United States for the gene BRCA1, which in certain cases is associated with breast cancer. The test is based on the protected property rights to this gene. Since then, each procedure to diagnose this gene must be paid for, and hypothetically this would also apply to the daughter of the woman whose gene was originally patented.

The scientific community shares the concerns of the public, though for other reasons. If companies can acquire ownership of genes or gene sequences, free access to the use of genetic knowledge is blocked for researchers. The more genomes that are deciphered, the more urgent the problem. A number of initiatives—like the recommendations of the British Nuffield Council on Bioethics, the Wellcome Trust, and the personal endeavors of Nobel Prize winners like John Sulton—massively advocate a policy of open access. But many questions of patentability remain open in the field of genomics. A fluid state driven by the dynamics of science is still in a process of stabilization. As in other cases, in the phase of controversy, the law gives visibility to disputed characteristics. As soon as a successful standardization (and legal stabilization) is achieved, the controversies themselves recede into the background. One of the most important tasks of the law is to accompany and shape the introduction of new technologies and to bring them into harmony with human existence. First it was labor law, which in the wake of industrialization and its inhuman working conditions, had to successively adjust the disjointed temporal-spatial work structure to human needs. Prohibitions and legal regulations led to mandatory standards that cannot be suspended by contracts. These stipulations ensure workplace safety, the surveillance of hygienic conditions, and the protection of the weak (workers) from the strong (employers). The standards contributed crucially to minimizing the risks of an era that venerated technological progress.[3]

Today, many people regard the subversive potential of biotechnologies to dissolve social relationships, identities, and traditional meanings as a threat. They expect the law to have a calming and stabilizing effect on the social order. The law has the power to issue prohibitions, but the societal environment prefers a restrained approach to prohibitions. In an age in which the legislature is increasingly replaced by voluntarily agreed on contractual conditions (often under the illusion that individual contracts are concluded between equal partners), people also expect the law to strengthen the ability of individuals to organize themselves. The law must contribute to the creation of new institutions that can intervene to mediate among and regulate between the individual, the state, and the globalized world. In short, the law must come to an arrangement with the new regime of governance.

The law, both as *lex* (legislation) and as *ius* (legal practice), promulgates limits or bans that, as in the case of the ban on human cloning and germline modification, are to be valid in the future, as well. But does it make sense to erect absolute limits? What can humanization mean if science continues to change what is defined as human? And as a pragmatic consideration, the question arises whether, however vehement today's controversies may be, tomorrow may see the acceptance of whatever people will regard as useful for themselves.

The law must thus be open to societal and cultural influences while being equally intent on preserving its own independence. If the law opens up to proceduralization, in which content is for the most part bracketed away because those involved are willing to regulate themselves and bind themselves contractually, then it is following a worldwide trend that we will encounter again in the advance of governance. In the case of Diane Blood, for instance, the law made use of a principle that was developed in the bioethics discourse. In place of the unanswerable question "Should dead men be permitted to become fathers?" comes the legal instrument of informed consent, as practiced in every hospital routinely before major interventions. The question is now "Do I, as a (male) patient, in case of my death, agree to the preservation of my sperm cells and to their use for a later IVF?" This question can be easily answered yes or no because it transforms a complicated, hypothetical matter into a familiar living will.

The law therefore may not allow itself to be made a tool of economic interests or yield to political interests, even if they are morally motivated (as in the case when the U.S. President's Council on Bioethics considered the creation of ethically neutral embryonic stem cells). The law cannot directly adopt either scientific evidence or scientific definitions and apply them normatively. It must carry out its own creative activity of lawmaking and legal practice. It must create the common reference valid for its area of jurisdiction, on which the legitimacy of lawmaking is based. As the fathers of the U.S. Constitution wrote, it must invoke those higher truths that are "self-evident."

The problem is that the self-evident truths that arise in different political and cultural contexts are not identical. This may be a reason for the astonishingly great differences found between the national legislations for the regulation of biotechnologies in otherwise comparable democracies.[4] Bioethical debates, too, are powerfully shaped by national traditions and history. So it is no wonder that the European Union has so far not been able to move past a few first steps toward harmonizing its members' laws pertaining to biotechnology, not to mention the deep ethical rifts that characterize every discussion of embryonic stem cell research.

Noteworthy are the different degrees of flexibility with which different legal systems react to and try to anticipate new biotechnological and biomedical developments. The common law system of the Anglo-American legal culture is based on countless individual cases that have been decided by the courts. Here, legal practice has precedence over legislation. Each individual case has its story to tell, and each one can become a precedent. Precedents seem more suitable for dealing in a juridically farsighted manner with broader lines of technological and societal development and for intervening with appropriate correctives. In common law, arguments confront counterarguments. This provides greater scope for differing viewpoints and favors a form of argumentation that develops further in the debate with the antagonist. It gives rise to a pragmatic legal realism that makes the outcome of the entire process more open. These characteristics give common law greater flexibility, whereas the continental European legal system is more systematic and thus more foreseeable but seems to proceed less pragmatically (and less realistically).

The renowned physicist Freeman Dyson has proposed an interesting and seductively simple suggestion for the future acceptance of biotechnologies.[5] His diagnosis is that what bothers people most about the biotechnologies is the centralization of this (like every other) technology, which throughout history has always meant a monopoly on control. To counter these feelings of being at the mercy of an uncontrollable power, the biotechnologies of the future should therefore take the path followed by the information and communications technologies. Biotechnologies should maintain their decentralized accessibility for the users. Only then will they be perceived as user-friendly and hence be taken for granted as a part of everyday life. Only if they are made more easily accessible and become available—for example, in the form of children's toys ("Design Your Genome!")—will their domestication be facilitated. But even Dyson concedes that their legal and ethical regulation is still an open question. We suspect that this regulation will come more easily to common law than to the continental European legal system.

THE NEW HUMAN TECHNOLOGY OF GOVERNANCE

The transition from the regime of government to that of governance—from centralized regimentation to a regulation in which the participants voluntarily submit to a common goal—was prepared by the rapid spread of information and communications technologies and the laws accompanying them. It resulted from a general upgrading of the value of information, for whose production, processing, storage, and further use new standards were created. Criteria like transparency and the accompanying instrument of the audit take high priority in managing the wealth of constantly newly generated information, its multiple possible connections, and its further use. Protection of privacy and many other things are thereby recognized as urgent. In regard to these problems, too, the emerging tendency is to shift potential solutions of existing conflicts to the procedural level as much as possible (proceduralization). This leads to a situation in which contents regarded as intractable and difficult to deal with are tacitly presupposed and suppressed in favor of rendering the procedure transparent. Beyond that, social relationships are being contractualized. The participants are increasingly willing to bind themselves contractually (instead of expecting the legislature to bind them), which is further promoted by the decentralization that has taken place. The new software systems bring a wave of standardization with them—benchmarking, specifying milestones, refining performance indicators, constantly changing recursive goals, and performing evaluations and continuous assessments. They have become indispensable instruments in the implementation of governance in the private and public domains.[6]

On the macro level, the relative retreat of the state confronted with the advance of the market has left empty spaces that are being filled with new norms and standards. Seen in this way, it is no coincidence that the large transnational corporations and multinational companies became the trailblazers of governance. They were the first to be released from the protectionism of the nation-states and had to reorient themselves in globally operating markets. The rules of governance that thus emerged have proven so successful that, in the course of the spread of the new public management, state actors have adopted them.

But the pressure on the classical model of government also grew from another side. Since the 1960s, when environmental problems first became a constant component of the political agenda, normative, moral, and value-oriented perspectives have entered the public discourse and are testing the ability of traditional democratic procedures to find solutions. Events with a high degree of public visibility—like climate change, nuclear reactor accidents, the outbreak of epidemics of uncertain origin, and other near-catastrophes—have become recurrent triggers for seeking new institutional arrangements. As an answer to the loss of trust in public institutions and technocratic expertise, mostly local experiments with participatory or deliberative democracy—like citizens' forums and consensus conferences—have arisen and have made the need for new forms of regulation evident.

These processes have led to a situation in which the state and politics, as well as business and science, have to consider a growing number of actors in various kinds of ways. Their participation in decision-making processes, the inclusion of multiple interests, and ethical and moral evaluations have become, according to Renate Mayntz, integral components of the "intentional regulation of collective matters in the state."[7] Governance can thus be understood as the management of interdependencies. It emphasizes the state's growing task of coordinating, which has joined, if not replaced, government as a process of political steering, promising increased effectiveness in fulfilling collective tasks. As vague as the term *governance* may be, one can filter out of the extensive literature common denominators that can clarify the concept and contribute to a better understanding of its rise.[8]

First, governance is based on a selective perspective. It emphasizes the ability to solve problems after common decisions have been made and assumes that solutions can be found for all conflicts. It is committed to a pragmatism that stipulates that involving many serves the interests of every individual. Aspects of power as political rule have no place in it. Power is eclipsed, reminiscent of the genre of earlier totalitarian utopias that sought to cleanse the world of social conflicts without being held accountable for the instruments of power thereby wielded. "The world is [has become] flat" not only for the advocates of globalization but also for those of governance.

Second, governance does justice to the fact that the actors to be involved are extremely heterogeneous. It assumes that diversity enhances the quality of decision making. However the actors may be organized and regardless of their legal status, the potential ability to influence decisions is recognized for all. Everyone can exert direct or indirect pressure. The world of governance thus consists of potential decision makers who are identical to those they seek to influence.

Third, decisions are no longer what they were under traditional forms of government. There is no clearly identifiable point in time when things are decided or a problematical matter is unambiguously arranged. The decision-making process becomes an iterative and open weighing and coordinating, a constantly fluctuating negotiating and adjusting through which compromises are found and interests are attuned to each other. The authorities responsible for regulating are required to engage in a constant dialog with all parties. Thus, the process inherently has a preliminary quality in relation to an open future. The principle of reversibility has also taken possession of governance.

Fourth, from a normative and democratic political standpoint, the concept of governance raises questions of how democracy is to be understood. The spectrum of norms that are involved in the greatest possible participation of the actors is accordingly diverse—state and international law, conventions and habits, soft law and good governance. Governments, nongovernmental organizations (NGOs), and industries have differing norms but have joined together in network governance. It is tacitly presumed that the exercise of governance already guarantees the democratic political quality of the outcome, since the securing of democratic legitimacy lies in the diversity of the participating actors. Measured against the standards of democratic representation, however, the democratic quality of governance proves to be problematical—all the more so because its arenas are often decoupled from the institutions of representative democracy.

Fifth, the engagement of civil society in the framework of the governance procedure has shown that *arguing* (deliberating toward finding consensus) and *bargaining* (negotiating compromises toward substantial realization of the participants' individual interests) are more important for decisions than voting in a system of majority rule. This is one reason that the arenas of governance are decoupling from the institutions of representative democracy.

Sixth, the ethics discourse and the struggle for shared values play an unexpectedly large role in the search for a common basis for the complex, often unprecedented negotiating processes. The lack of a generally obligatory framework—or at best a loosely obligatory and abstract ethical framework—is compensated by opening a wide space of interpretation. Seemingly incompatible positions and valuations are to be rendered capable of compromise through processes of argumentation and negotiation, which is possible only on binding ethical foundations.

Seventh, governance creates a public realm in which all participants come together as stakeholders. Independent of their status as minorities and of the identities of the majorities, all members of civil society are admitted. Everyone who wants to speak is to be heard. A kind of pluralistic, inclusive community that forms temporary alliances thereby arises, but due to the lack of common values, it tends to be restricted to the smallest common denominator.

While some (like Bruno Latour) see nothing but a "sad dream" in governance, others observe a further development from "governance through risks" to a "governance through uncertainty."[9] In the 1970s, the aim was first to stabilize and regulate the risks associated with the dawn of genetic engineering. Since the early 1990s, the risks have increasingly shifted from the outside to the inside—into the body, its tissues, its cells, and its gene sequences—and new moral dilemmas and ethical ambivalences have emerged. The risk management that focused on the risks of technologies that exerted effects from the outside had no answers at hand.

The scientific or political uncertainty that arose from the lack of clear criteria for evaluating a number of biomedical practices became the real current theme. This uncertainty encouraged the mobilization of the people concerned, who began to use their own view of things to fill the empty spaces in which there were no official answers. They told their personal stories in a language that offered scope for emotions, subjectivity, and themes like suffering and pity. This emotional discourse, which was not always free of bathos, turned out to be compatible with a "politics of ethics and morals." Although the risks have lost none of their relevance, the emphasis on uncertainty is more than the admission of a lack of secured knowledge. It is another attempt to deal creatively with the diversity of stakeholders' experience. The necessity to introduce another standardization thereby shifts to

bioethics, which has become one of the most important instruments in the governance of the life sciences.

In *Playing God? Human Genetic Engineering and the Rationalization of Public Bioethical Debate*, John Hyde Evans reconstructs the
rise and institutionalization of bioethics.[10] Invoking Max Weber,
Evans speaks of a continuous "dilution" of the issues with which
bioethics is concerned. In its beginnings, bioethics was a "dense"
debate centered on both the aims and the means required to
achieve them under conditions of growing democratic participation. Soon the emphasis shifted from public control through
legislation to bureaucratically organized commissions with con-
sulting experts. Initially, theologians and philosophers could be
found on these ethics commissions, but later the commissions'
composition and the style of advising changed. Bioethics professionalized and expanded its international networks. Its context,
too, became formally organized in a thoroughly bureaucratic and
highly professionalized way. The discussion shifted away from
the objectives and concentrated on the means. Principalism came
to dominate consultations:

Through the articulation and application . . . of principles, standardized
rules were established that permit the translation of differing moral principles into a common metric that—usually on a cost-benefit basis—offers choices and decisions. If the principles are to function efficiently,
together they must form a system that permits weighing, predicting, and
predictability: the qualities of a currency.[11]

The underlying principles have a utilitarian function that aims
to permit certain moral arguments and to exclude others. After

viewing the massively grown literature on bioethics, Salter and Salter come to the conclusion that the political value and function of bioethics make it an effective "currency of a global moral economy."[12] Its principles are flexible enough and have proven their ability to adjust successfully to new circumstances. Bioethics itself has thus become a political means for negotiating and even exchanging values. This has made it a neutral technology of normalization and legitimization. Where the authority of science alone no longer suffices, bioethics has joined it as a state-sanctioned authority. Bioethics has the right instruments to translate moral evaluations into a common language characterized by formal rationality and efficiency. Like every expert's language, bioethics is free of emotions.

For pluralistic societies, it is indispensable that the methods of bioethics be presented as inclusive. They offer room for legitimate differences to which bioethical principles like autonomy, benevolence, the avoidance of unnecessary suffering, and justice can be applied. As a currency for a transnationally functioning moral economy, bioethics must be transferable to other cultural contexts. In the ideal case, it offers itself as a neutral technology that always has a correct solution at hand, independent of persons, time, and place.[13]

The rise and relevance of bioethics are indeed as remarkable as their present indispensability for the life sciences. Their political function as an instrument of legitimization is hardly lessened by arguments for the "soul of bioethics" that aim at the invocation of contents and goals (and not just of the means to achieve the latter). Their guarantee certificate for the legitimate circulation of egg cells, genes, tissues, and other somatic components in global biocapitalism is too important. After all, they are the precondition for the emergence of what John Durant and colleagues call a "biotechnological complex," in which the life sciences can ally with powerful economic interests.[14] Nevertheless, in the future, the biotechnological complex will have to deal with the tension that has grown between a highly professionalized, unemotional, official *bioethics* and a subjective *biomorality* that draws its strength from the experience of affected persons and from public discourse's openness to emotions.

But a return to the "dense" ethics discourse (that is, to a discussion that also addresses ends and not solely means) would completely overstrain bioethics—and the law. It would burden them with a task that neither politics nor religion has so far managed to fulfill—erecting a new communal project of society or even reshaping our pluralistically fragmented life on a foundation of common values. The success that the three human technologies examined here have had so far in stabilizing the social order is based mostly on a relatively successful standardization. It distances itself from pinning down common goals and instead creates procedures that permit advances on many different paths. It could be said to have no answer to the question of where we actually want to go. But this does not hinder us, as a society of modern individuals, from moving forward.

A similar approach plays a decisive role in synthetic biology, which is in the process of turning some of the life sciences into engineering science. In both cases, the aim is to create standards that permit a change—a reshaping—of forms of life. A deeper convergence of the molecular age is thereby revealed. Human technologies that have reached a certain degree of societal maturity are converging with a biology that is open to societal goal-setting, to taking legal and ethical limitations into account from the beginning, and to including them in its design. Common to both is that they are complex systems that should be disassembled into their component parts and reassembled. The synthetic design of life orients itself toward the social design of society. But both will have to confront the questions and demands resulting from the uniqueness of life and from the unique individual experience each of us has. Flexible procedures, however fluid they may be, need to be harmonized individually and on the societal level with the density of human experience. The future will show whether these complex biological, technological, and social systems have an architecture that is stable enough to hold up to changing and partially unforeseeable demands.

6 SYNTHESIZING THE FUTURE

THEME 6   The efficient arrangement of standards that enable the measurement and unification of molecular life finds its correlate in standards of socially, politically, and ethically responsible behavior. This convergence opens a view of a future in which the standardization of life is advanced explicitly in this double sense. One of the most current areas of research, synthetic biology, has set itself the goal of rearranging life by designing its components. But to this end, these components must first be made capable of standardization. In this chapter, we show how synthetic biology is developing beyond the ad hoc use of technological procedures into a systematically pursued undertaking. One example of this is the complete sequencing of the microbial diversity of the ocean. Starting from the parts acquired in this way, artificial components will be produced that can be reassembled for specific functions. Standards and an accessible register of standards play a crucial role in this. To conclude, we use an astonishing example to elucidate one of the characteristic traits of synthetic biology—the coupling of biological and social standards (in this concrete case, with moral values). The syntheticization of the future raises *design* to the leading principle. On the one hand, this is a logical extension of the engineering sciences and their ideal of designing. On the other hand, this makes it clear that designing standards is much more than a question of technology.

THE SEARCH FOR THE COMPONENTS OF LIFE: THE DNA OF THE OCEAN

We encountered the science entrepreneur Craig Venter earlier in a discussion about high-throughput sequencing technology. Sailing around the world with his team to take samples of the "genome of the ocean" was the next challenge for his research empire. Synthetic biology—producing biological systems artificially—is a rising and expanding discipline with its own research institutes, curricula, and specialized journals in which researchers in biology, chemistry, informatics, and engineering sciences come together with the defined goal of redesigning existing biological systems and constructing new biological building blocks, instruments, and systems for useful purposes. Although synthetic biology is still in its beginnings, it also has its prehistory. Gene technology has been practiced for three decades. In the first days after the successful recombination of DNA in 1978, two of its pioneers announced that the possibility of putting together pieces of DNA in new ways by plan would usher in a "new era of synthetic biology" in which "new arrangements of genes can be built and evaluated."[1] But the leading researchers in synthetic biology are by no means merely pursuing a seamless continuation of gene technology as practiced in laboratories all over the world. Rather, the ambition of synthetic biology is to ovecome the current research mode, in which organisms are manipulated through expensive and labor-intensive ad hoc experiments, into a new phase in which biological systems can be assembled "off-the-shelf" from standardized parts.

Until now, the successful arranging of *in vivo* systems and technologically producing simple living organisms (above all, bacteria and yeast with predetermined characteristics) have been the expensive result of ingenious research plans. The rearrangement of life from its beginnings is far from being a routine procedure in gene technology. Students of electrical engineering learn very early how to build a circuit out of standard parts. The concepts underlying the standardized components are well understood, and the parts themselves are readily available at all times. By contrast, a bioengineer would have a difficult time compiling a catalog of reliable component parts whose meaning and function are stable enough to put them together as a new piece of life. A trivial explanation of the difference would be to invoke the complexity of life. Our understanding of biological systems is still insufficient for a technological access that would make rearrangement routine. But from their own perspective, the ambitious representatives of synthetic biology are right when they argue that we cannot know anything before we have tried to produce it. This stance is based on a Cartesian understanding of the organism as a machine and on the assumption that we do not understand an organism until we can construct it from the bottom up. For synthetic biologists, "biotechnology remains a problem of research because we have not yet invented and employed any basic technologies to turn it into a technical problem; biotechnology remains complex because we have not so far made it simple."[2]

Craig Venter has made this radical simplification his program. The expedition of the *Sorcerer II* is a perfect example of synthetic biology. Fully equipped with the latest molecular-biological techware, the *Sorcerer II* sailed down the coast of North America, through the Panama Canal, and on through the Pacific and Indian Oceans. At intervals of about 200 miles, it took 200 liters of seawater as a sample of the diversity of the aquatic environment. These seawater samples were then pumped through filters, and the DNA of its microorganisms was extracted and sequenced in record time. These DNA sequences taken from the oceans thus represent the DNA of the global ocean. Only afterward are ingenious algorithms used to put together the sequenced fragments as the individual genomes of the various microorganisms that were pumped out of the water and through the filters. But the primary goal remains the sequencing of the ocean's DNA as such, the aggregate of the individual parts that make up its microbial life. The real goal of the project is the genetic identification of these components. Significantly, the authors describe their work as "gene-centered" to distinguish it from the "organism-centered" procedures of metagenomics, in which whole environments are registered by the simultaneous and complete sequencing of their microbial diversity.[3] Species and organisms move into the background, while their constituent components—the DNA fragments that encode these entities—move into the foreground.

Comparing the journey of the *Sorcerer II* with Charles Darwin's voyage on the *Beagle* reveals far-reaching changes in the questions posed during the two expeditions. In the autobiography that he wrote at the age of seventy-two, Darwin writes that, as a child, he already felt a "passion for collecting which leads a man to be a systematic naturalist, a virtuoso, or a miser."[4] Darwin observed and collected organisms, thought about how species develop, and drew from his observations conclusions for a model that explained the origin of the diversity of species. (At the end of his writings, he complains that his mind has become like a "machine" that has to compress an enormous collection of facts into general laws.) Less than 200 years later, the researchers on board the *Sorcerer II* are not primarily interested in observing and collecting organisms, even if they do archive them and even if the organisms' identification is one of the undisputed successes of the project. They focus on the diversity of the parts gained by means of the sequencing. Their gaze is directed into the interior of the microorganisms, at thousands of new proteins and the unexpected forms that many of them display. The spectacular success of the expedition lies in detailed and precise identification. It is itself the result of knowledge about the informational content of the DNA and of the ease with which the samples could be extracted and analyzed.

This success is all the more valuable because it has to do with the ocean, the archetype of everything that is large—large in volume, life, and time. A catalog of all the organisms of the ocean and their specific environments remains beyond human grasp. Nonetheless, a catalog of the ocean's genome is compatible with the spirit of the times in today's science. But there is another striking difference between Darwin's and Venter's voyages of discovery. Darwin's journey was devoted primarily to observation and collecting. Crates and crates of collected objects were sent to London to help researchers take an inventory of nature and examine the connections between its creatures. From the first day, Venter's undertaking was a trip with the goal of intervening in nature. The oceans were to be searched for all its microbes (in relation to the earth, one would speak of "prospecting"). They are the raw material of life, whose wealth consists in its biological parts and their unimaginable diversity. These parts await their further use in the project of synthetic biology, which is open to the future. They are the anticipation of the potential that is to be tapped.

STANDARDS AND REGISTRY

But what exactly are the usable biological parts that consist of genes, enzymes, and metabolic pathways? What does it mean to redefine them in the context of synthetic biology, and what follows from their identification? Perhaps it is best to ask the synthetic biologists themselves how they see it. The Registry of Standard Biological Parts housed at the Massachusetts Institute of Technology is one of the furthest-developed attempts at determining how units of biological functions can be brought together as standards.[5] And standardization is indeed the core of synthetic biology. The "central dogma" of molecular biology—the directed flow of information from DNA to RNA, to proteins and phenotypes—is not only a model for interpretation that is historically based on the "information turn" that led in the 1960s and 1970s to the idea of the genome as a program. For synthetic biology, this "dogma" *defines* the operations of biological systems. These operations must be describable in standardized form, which is the precondition for making them measurable, comparable, and, even more important, exchangeable. This precondition has not yet been fulfilled. The deficit has to do with the way that molecular biology has developed in the last few decades. Up to now, the biological parts have been investigated in completely different contexts in the widest variety of organisms, depending on the goals toward which the individual laboratories conducted experiments. The existing results are thus due to individual scientific successes. They are not the result of a planned, joint research strategy. They stand for (artistic) craftsmanship but not yet for industrial manufacturing and use.

The goal of synthetic biology is to make prefabricated parts out of biological components and then to use these parts to design new living systems. The behavior of the components of the new system will not be reliably predictable to any degree until they have been produced on the basis of known qualities and generally valid standards—as parts that have been defined by standard measurements in standardized experiments and that have therefore become standards themselves, around which the design of the new life can be organized. The Registry of Standard Biological Parts is a first attempt to fill the shelves of life with prefabricated parts. But other shelves are waiting with molecular prefabricated parts, as a glimpse into the molecular cuisine reveals. Chefs who like to experiment, supported by researchers who like to experiment, are pleasing their guests' palates with culinary novelties. Another example is provided by scent manufacturers. Each year, the market sees 600 new perfumes that are based on the production (and patenting) of synthetic molecules that offer new combinations of olfactory experiences for our sense of smell, which has atrophied in the course of the evolution. The vibrant discussion about standards in synthetic biology now already comprises the discussion about legal and ethical standards, although these, like the acquisition and construction of biological building blocks, are still fragmented. Currently, researchers and research centers must negotiate the legal framework conditions for the use of each of the individual genes, cells, and animals that they have to share with others, using a standard instrument, the Material Transfer Agreement. But these standard agreements contain only a few real standards. Each of these agreements reflects the biography of the biological part in question. Therefore, the same part can wander through the various institutions, affected by very different agreements.

The problem of standardization becomes all the more acute with the rapid advance in the technology of synthesizing very long molecules. Until recently, it was difficult to synthesize DNA chemically. It was not possible simply to order thousands and ten of thousands of DNA "letters" and await their delivery. Rather, they had to be cut out of existing molecules and glued together again. But advances in chemical synthesis and the polymerase chain reaction (PCR) will make it possible in the foreseeable future to order entire genomes. How will patents and intellectual property be regulated? Will they become part of a copyright that reduces life entirely to the text and protects the engineer's invention as if it were a song or a literary work?

An instructive example of the dynamic development in this field comes from the International Genetically Engineered Machine Competition, which was set up at MIT in 2007 for students of synthetic biology. The participating groups receive a set of biological prefabricated parts. During the summer, they work at their home institutions to put these prefabricated parts—along with new parts that they design for building biological systems—to work in living cells. In the fall, the best parts and systems are awarded prizes. The company Geneart, which specializes in the chemical synthesis of DNA and whose slogan is "The Gene of Your Choice," was the leading sponsor. It offered the teams of students substantial price discounts when ordering the first 100,000 letters of DNA.

Even if the production of a mail-order catalog of biological prefab parts is still in its infancy, bacteria and yeasts are already being synthetically manufactured (for example, to be used to remove toxic waste or to synthetically produce artemesinin, thereby reducing the costs of effective antimalaria medications by 90 percent). Nor is it surprising that synthetic biology soon came into the sights of the biosecurity experts, since pathogenic germs or organisms can be produced in the same way. What threat this already or potentially poses and what countermeasures should be taken depends on more than the degree of achieved standardization.

When the bioengineer Vincenzo Tagliasco decided not to publish another edition of his 1999 *Dizionario degli esseri umani fantastici e artificiali* (Dictionary of Fantastic and Artificial Human Beings), one of his considerations was that a long tradition of fantasy figures—created by scientists like J. D. Bernal and J. B. S. Haldane through Hans Moravec to cult films like *Blade Runner* and many science fiction authors—would soon be outmoded by the already existing artifacts of imminent technologies. Our bodies, long since accustomed to prostheses and implanted organs, will perhaps soon come to know the results of stem cell research (whose technologies derive from those of animal breeding, as Dolly's immediate predecessors Megan and Morag show). But synthetic biology does not aim to extend or implement science fiction. Its impetus comes from the engineering sciences, except that its building blocks consist not of inorganic matter but of life itself. For a long time, science fiction and its fables of a technological era fulfilled an important function: reading fictions was to help us think about the future. The literary genre of science fiction demands an alarming tone, the exciting feeling of facing a danger, and the arousing presentation of the possibilities of a future that no longer has any place for humans and human values. Its warnings are inventive, often witty, sometimes salutary, but hardly calming reading. But science fiction has never shied from asking the big questions or from projecting into the future what later became facts that mold our everyday lives. Perhaps it will keep us alert to the questions that we must continue to ask if we want to create a human context for the developments resulting from the convergence between biological and social standards,[6] just as literature of every kind opens up a space in which we can gather imaginary experience that expands our repertoire of experience and experiments.

THE FUSION OF BIOLOGICAL AND MORAL STANDARDS

The attempt to make biology a subdiscipline of the engineering sciences is merely a logical development of a number of procedures for creating new life that existed before the planned design of biological material could be conceived. Cloning and the rise of stem cell research over the last ten years are the paradigm of this development. Dolly the sheep is indeed the embodiment of a new procedure for bringing mammals into the world. It may not be extremely synthetic, since it is based on a merely unconventional method of mixing components (an egg cell and a somatic cell nucleus) that cannot yet be reduced to standardized prefabricated parts. But it lends great visibility to a procedure that synthesizes life from a material (somatic cells) that at the time was not considered suitable for this purpose. If the core of synthetic biology and its mode of thinking consists of producing new combinations of living material and in designing new biological functions for it, then we have long since entered the age of synthetic biology.

The ultimate desire of stem cell researchers is to be able to reprogram any cell of the body to develop into any other kind of cell. From the beginning, this was the vision that stood behind therapeutic cloning. The attempt to use somatic cell nucleus transfer aimed to produce not a clone of the patient but a cloned embryo from which embryonic stem cells tailored to the patient could be won and then differentiated into the needed type of cell. The advantage is clear: since the transplant consists of the patient's own "backup" cells, her immune system need not be suppressed. But the low efficiency in somatic cell nucleus transfer, the lack of human egg cells, the ethical problems associated with acquiring them, and a widespread, if inconsistent, moral resistance to cloning have stimulated the search for alternative procedures. This is why an originally scientific debate about the priorities and differences between embryonic and adult stem cells turned into a political confrontation. The search for alternatives recently spurred Japanese researchers to use adult skin cells to create pluripotent cells that share the primary characteristics of embryonic stem cells.[7]

Especially worthy of mention among the alternative sources for ethically unobjectionable embryonic stem cells is altered nuclear transfer (ANT). It was suggested by U.S. President George W. Bush's Council on Bioethics in 2004. The ethicist William Hurlbut said that ANT would make it possible to use the desired effect of cloning and of embryonic stem cell technology while simultaneously avoiding the ethical misgivings that accompany the destruction of human embryos.[8] The idea was simple and ambitious at the same time. If it were possible to deactivate a gene in the somatic cell nucleus before transferring it to the enucleated egg cell—a gene essential for further development—then the entity thereby created (called ANTity)[9] would lack the traits and capabilities of a human embryo. In biological and moral respects, it would be more similar to the only partial developmental potential of a tissue or cell culture.[10] The entity thereby created would be so crippled that it could never grow into a fully developed human being. Its use would therefore be morally unobjectionable. The crippling would have to be so carefully controlled that this ANTity could reach but not go beyond the blastocyst stage, which is the precondition for gaining embryonic stem cells.

Leading scientists at MIT took up this suggestion and literally brought it to life. The journal *Nature* published the report on the successful production of the new biological entity, an early stage of life that cannot be implanted in the uterus because an essential gene has been deactivated.[11] But many researchers remain skeptical about an approach that is supposed to conjure up moral ontologies from the wizard's cauldron of biotechnology. Michael Gazzaniga, a respected neurobiologist and a member of the President's Council, was against ANT. He complained that we normally coin a word to describe a biological phenomenon but that here we seemed to tinker with a biological phenomenon to make it fit the meaning of a word. But that was the intention. If representation and intervention blend together, then words determine both the meaning and the existence of new forms of life. Moral convictions mix with the most sophisticated biotechnologies. The discoveries and inventions made in laboratories are thus increasingly in a position to carry within them a double certificate of origin—a scientific one and a moral one.

Beyond the specific arguments, in the heat of the debate about ANT, the actual significance of this trajectory from the Council on Bioethics through MIT's high-tech laboratories to publication in *Nature* went mostly unnoticed. Never before was the relationship between science and politics this intimate and at the same time public. Deadlocked moral or religious views on the value of human life and on the definition of its beginning had dictated the creation of a biotechnological artifact—a "human creation for human purposes," as Hurlbut described it. A biotechnological creature came into being that was tailor-made to mediate between religious convictions and the pressure for scientific and technological innovations. When it came to the question of which gene should be deactivated to make this biological artifact unproblematic for all sides, the primacy of politics over science could not be doubted. Exactly which gene or combination of genes should be deactivated depended on the level of developmental disturbance that is necessary to fulfill the moral criteria of this project. And so we have returned to standards—this time, the moral standards that fuse with scientific ones.

To deactivate the gene that is necessary to form the placenta, ANT employs a technology called ribonucleic acid interference (RNAi). The gene is not physically eliminated, but a small RNA molecule prevents its expression. What provides the certainty that this tiny inhibitory RNA molecule does its work reliably in such a complex system and that the target protein is really no longer present? If the entire project is based on the assumption that using this embryo-like entity is ethically neutral only if this protein is lacking, how can we ensure that this is really the case? And if, despite the suppression of the expression of the corresponding gene, a small amount of the protein remains, how large a remnant is the threshold at which our ethical obligations are no longer fulfilled? The answer is that it depends on the standards.

In one of the leading journals, the *New England Journal of Medicine*, a heated debate broke out about how the absence of this protein and thus the moral worthlessness of the created entity could be irreproachably determined.[12] To this purpose, it was suggested that the small RNA molecule that leads to the destruction of the cell protein be tagged with a glowing marker, the green fluorescent protein (GFP). Its activity is easy to measure, due to the green light it emits. Since the expression of GFP is tied to that of the tiny RNA molecules, we know that the greener the cell, the more of such inhibitory RNA molecules it produces. Consequently, we can say that the greenest cells are also the morally least objectionable. In the twenty-year career of GFP, which is used in all laboratories around the world to study gene and protein function, this is the first time that the power of its glow indicates—indirectly but all the more intensely—a moral value. The standard measurement in the laboratory—the glow of a green signal—is raised to a standard of moral value, and vice versa.

7 THE FUTURE AS SYNTHESIS

And Bhima said, "Tell me how many epochs there are and what happens in them, and in what state the law, profit and pleasure, greatness and power, existence and death are."

—*Mahabharata,* translated by Geoffrey C. Bowker

THEME 7 The conviction that we stand before an epochal breakthrough with revolutionary possibilities is nothing new. It accompanies every technological vision. Despite novelty's inherent fascination and its supposed discontinuity, we must not lose sight of the power to persist and (as David Edgerton calls it) the "shock of the old." The continuity of societal developments should not be underestimated. In this last chapter, we attempt to synthesize the relationships among three reference points—a kind of triangulation among science, individuals, and institutions—to open the view onto a possible shaping of the future. First, there is an increasingly globally operating science whose form of organization is rapidly changing. A scientific superorganism is emerging whose decentralized parts are networked with each other. The second observation has to do with the experiences of individuals and the question of what it means for them to find themselves willingly or unwillingly connected on the molecular level. Their identity and (multiple) affiliations are being determined anew, genetically and socially. Creative individuals who take their destiny in their own hand are one possible answer to this fluid situation. The third reference point is institutions. Their task is to take social practices whose official status is still controversial or indeterminate and stabilize them in such a way that a pluralistic living together is possible. They have to offer a legitimate free space that allows experimentation with new biological forms of life and of living together socially. Institutions should empathetically accompany individuals who are prepared for both. In the interactions among these three reference points, space for the primacy of the political is created. It redefines the common good and options for society and its individuals "beyond the arbitrary polytheism of social values and individual opinions and preferences" (as Laurent Thévenot has noted).

PRODUCING THE LATENT FUTURE: A SUPERORGANISM EMERGES

The decoding of hundreds of genomes and their systematic comparison has given rise to millions of gene sequences that have been registered, processed, and archived in comprehensive libraries and maps. This work—with these data, pieces of tissue and blood samples, gene sequences, stem cells, and their lines—presupposes the construction of carefully chosen procedures. Each step is noted in a protocol. The technological standards are supplemented with ethical and legal ones (what rights are owned by whom, under what circumstances organisms may be exchanged, and what should be placed in public repositories). Thus, the article that was published in *Nature* in November 2007[1] on the occasion of the successful cloning of primate stem cells was flanked by a position statement from an international reviewer group. It aimed to ensure that no fraudulent manipulation was involved and that the created embryos were indeed genetically identical to the organism from whose cells they originated.

Mostly unnoticed by the public, the rising tide of data and of its numerical processing has led to far-reaching changes in scientific work processes and forms of organization. For example, a mathematicization of biology has ensued, and developments are moving it increasingly toward the engineering sciences. But the effects of these organizational changes are much broader than the as yet unknown epistemic challenge of the analysis of a large number of genomes whose building blocks are purposefully reassembled and will thus lead to a redesign of life. As discussed in chapter 3, the dinner with Craig Venter and the founder of Google reminds us that the organizational arrangement of contemporary life sciences is intimately embedded in the flow of capital and information that drives the global market. This dinner, with its far-reaching consequences, casts a spotlight on the emergence of a scientific-organizational superorganism. We have encountered it several times in this book. Each time, the point was to domesticate the central characteristic of life—its *variability*—and to employ it in a useful way for science and society.

To this end, bearers of the genetic material to be compared must be continuously recruited because only after genetic information from a sufficient number of individuals has been pooled and compared can one gain relevant information for individuals. Whole populations or subgroups thereby become scattered depots of the new and precious raw material of genetic variability. The pool of genetic variability unites these resources and domesticates them at the same time. For example, on the homepage of a pertinent research institute, we can read the announcement that it now wants to use "selected human resources for the hunt for disease genes." This resource could be the Icelandic population, which, as approved in parliamentary debates, has sold its DNA to a company. Or it could be the users of Google who want to inform themselves about a specific genetic risk, thereby becoming the consumers of a genetic test offered on the Internet. In both cases, the visibility of the genes lends them epistemic significance and social relevance. We understand their biological function, including the metaphorical baggage accompanying them, by means of the digital mediation that prepares genes made visible in this way for the information highway and thus for the world market. High-tech biocompanies, private and public funders, patients, and all who seek preventive control over their fate encounter each other in a newly emerging market, the environment in which scientific work takes place today and to which its organizational forms must adapt. Sooner or later, whether or not we know and want it, we will all become parts of this superorganism.

It could be objected that this is nothing new. Through and for their new administrations, emerging nation-states collected statistically relevant material, initially about young men who were eligible for military service and soon about the entire population. The great vaccination campaigns and community-wide hygienic measures to stem epidemics were based on a registration of the populace that was as complete and comprehensive as possible. Here again, we see the lines of a historical continuity, with the difference that the information gathered today encompasses literally our "molecular soul" and nothing less than Craig Venter's "genome of the ocean." The genetic infrastructure projects arising today operate with an interior gaze—and do so globally.

Biobanks and bioarchives are places where what seems to be a latent future is created today. The categorization of tissues, blood samples, and gene sequences extracted from different organisms in widely separated places at different times makes the scientific exploration of life a mammoth-scale project. It also includes an evolutionary past with its reconstructed lines of descent stretching back millions of years. A huge, genetically based inventory of life is emerging that includes projects like the barcoding of life. A new method of automatic identification of unique sequences that act as the barcode of life will produce a complete catalog of the species.

But as important as the evolutionary past may be, the gaze is directed toward the future. The questions that such projects will answer have not even been posed yet. The purposes for which the archiving, classifying, standardizing, and comparing are done are not yet known. The information contained in the "selected human (or other organic) resources" must be arranged in such a way that it is suitable for all future questions and purposes. So the question is not what is known about a gene, a genome, a worm, or a human being but what knowledge about them will be available from now on, *if need for it arises in the future*. The scientific and biopolitical superorganism therefore acquires the characteristic trait that, from the very beginning, has driven modern biology to corresponding efficacy and power—the construction of experimental systems capable of coming up with answers to questions that research cannot yet articulate in full clarity. What François Jacob said about the arrangements of the smallest complete working units of research in experimental systems—that they were "machinery for the production of the future"—is all the more true for the superorganism now coming into being.[2]

Other large-scale projects were dependent on large amounts of data and their preparation by specific scientific practices—like modeling, information management, and the construction of organizational infrastructures. A look at the history of the earth sciences, climate research, and the exploration of biodiversity permits a clear conclusion: we organize information in harmony with the way that we organize the world.[3] The way that the organization of information mediates between the world and our image of the world is characterized by further developing both. This happens because the different interpretations of the past must be continuously adjusted to the questions of the present. The present, in turn, must be prepared in such a way that it remains open for possible future questions. What Bowker called "practices of memory" are in reality practices that serve to create a latent future and to give visibility to something that is already inherently possible but not yet existent. This presupposes an enormous coordination achievement based on the standardization of innumerable, locally distributed social practices. The social form of organization thereby adjusts to the objects of scientific inquiry and becomes their mirror image. The scientific superorganism that is composed of worldwide consortiums, networks, companies, and universities with their private and public modes of financing, management, and governance has become astonishingly similar to the object of investigation—genetic organization.

THE SUCCESSES OF STANDARDIZATION

One of the most important and most efficient mechanisms of coordinating or coproducing the superorganism with its inner and outer world is standardization. It is indispensable for biomedical innovation. The effectiveness of medicines must be reproducible. The extraction of biological samples, the management of the huge amounts of data that result from it, and the manufacturing methods for tissues and cells from many local practices must be carried out in compliance with generally accepted principles. It's revealing that leading specialized journals are publishing more and more articles that suggest new standards for making sense of the data and giving it order. This is true even of the names that the genes are given. As a Japanese scientist once jokingly remarked during a conference, scientists are more willing to share their toothbrush than the names they give to genes. So one and the same gene often receives several names, each reflecting its epistemic origin or personal preferences. *Ad hoc* commissions follow this with periodic attempts to pin down new nomenclatures and to muster the scientific community's agreement on them. If the new biology of networks now searches through the increasingly growing databases of publications to track down relevant interactions (on the assumption that two genes belong together functionally if their names are found in the same publication), the task of standardization is anything but trivial. It is directly confronted with the *Zeitgeist* of informatics that characterizes contemporary biology.

The need for technical standardization is complemented by the need for legal and bioethical standardization. The bioethical discourse on the use of human body parts insisted very early on two limitations: the donation must be voluntary, and it must be anonymous. Voluntary donation aims to ensure that donations are not subject to commercial considerations. Anonymity creates a space that regulates the behavior of all those involved and the technical procedure. The implementation of these principles led to a lasting stabilization.[4] The experiences of affected persons and their relatives in the course of such standardization and their articulation by patient groups and other interest groups is initially often expressed in protest, controversy, and dissent.

The consolidation of social problems is thus not present from the beginning but is accompanied by public discussion and their conflicts. The chosen standards must be revised and improved in an iterative process of testing and quality control. No standard remains unchanged over time. Nevertheless, this kind of production of knowledge and consensus is made efficient and simultaneously economically sustainable by the dimensions and operating range of standardization. Scientific progress is not a linear process (and never was) but depends on new questions that often alter the way that science works. Standardization is a way of fruitfully preparing the ground of data—with its newly discovered, precious raw materials—for new scientific questions.

The global model of organization of contemporary science includes many geographically scattered actors in institutions in a wide variety of ways. In the United States, private companies that use knowledge to make profits are entering into new alliances with groups that advocate for patients' rights. In countries like Britain, where large charitable organizations play a role in shaping basic research and often spend more than the state, the institutional climate arising among medical experts, health authorities, and citizens is different from that where the state traditionally plays a greater role. And so, in different ways, the market, the state, or intermediary institutions operate as a unifying force.

The production of knowledge has therefore become both pluralistic and global. Networks are strengthened by the global circulation and exchange of information, samples, and cells and by the outsourcing of technical work steps. The dynamic of the superorganism requires the continuous synchronization of the newest technological procedures, revised bioethical principles, and legal regulatory instructions with their governance. The emerging superorganism has thus become a global variant of mode 2 knowledge production.[5] The participating actors from a wide variety of institutional environments of privately and publicly financed science come together to work in contractually agreed-to configurations on problems that, at bottom, have to do with jointly decided scientific questions and their potential use. As in all large organisms in which knowledge, money, and power are concentrated, the central question is: Who controls what happens? How are unavoidable financial conflicts of interest to be regulated, considering the inherent interconnections in the superorganism? How are values to be standardized in such a way that they can become an accepted currency of a global moral economy, as is shown in chapter 4? In other words, if the success of standardization is so indisputably great for the production of knowledge and for the moral and real economy, then where is the problem? Why does a continuous consolidation lead not to general mollification but to ever-new dislocations in the moral and political landscape and in the private emotional household?

STANDARDIZATION IN TIME AND SPACE

Standardization is a process that plays out in time. After a phase of protest and defensiveness against what is initially considered unthinkable, often a phase of calm and consolidation sets in. Let us recall, for example, the public outcry triggered by the birth of Louise Brown, the first "test-tube baby," in 1979. Since then, more than two million children have been born as a result of IVF. But not every novelty is accepted, even if the history of innovation—an always selectively narrated tale—shows that an initial phase of contestation is often followed by one of domestication, adoption, and adaptation. This story is selective because it blanks out the forgotten, failed, and dead-end innovations. So while it is true that initially unfamiliar social practices move into the background and are taken for granted in the consolidation phase, by far not everything is standardized that could be. And as we have already mentioned, standards change over time.

Another reason for their limited range is that their area of validity is both politically, geographically, and culturally molded. The standards and frames of reference of the different national legislations still differ as substantially as national ethical guidelines do. Institutions differ in their practices and cultural influence and in their proximity or distance to politics, business, and citizens. The spectrum of legislation stretches from liberal through restrictive to repressive. Some countries, like Britain, have their own longstanding traditions in which technological innovations and a kind of moral calculation converge within a liberal framework. Other countries proclaim certain values, invoking absolute, uncompromising standards that are "nonnegotiable." Bioethics also knows national differences. As successful as bioethics might be as a global currency in keeping controversies contained, neither it nor religion can claim moral universality (and bioethics is meanwhile professional enough to avoid even trying to do so).

A pluralistic society can do justice to these circumstances in the political and everyday senses of living together. The individual does not live in a social vacuum. His or her options and decisions are always shaped culturally, economically, and socially. Private morals can never be equated with public morality. What ultimately counts is public morality. Lee Silver, who investigated worldwide differences in cultural-spiritual attitudes toward genetically modified organisms (GMO) and the cloning of embryos for research purposes, encountered fewer reservations toward them in Asian countries, with their Hindu and Buddhist traditions.[6] Where there is no single creator God but many gods or none, there is also no master plan for the universe toward which the social order must be oriented. The spirit is considered immortal, and individual virtues—karma—determine what will happen in one's next life. Thus, options come in "packets": they are culturally presorted and unite moral values and social and economic opportunities in various configurations when encountering the offers of the life sciences. The superorganism depends on standardization, but the complexity is so great that the individual retains a great deal of freedom. If there is one thing that standards are not, *they are not homogeneous, even if they homogenize.* Like variability in life, they are inscribed with an irreducible diversity that should be respected.

REVERSING TIME: REVERSIBILITY AND THE SOCIETY OF WEAK EGOS

We asked ourselves whether life outside of standards—multiple, plural standards—is even imaginable. Our answer is no. But if that is correct, then there are at least two ways for the standardization of life to proceed in the double, simultaneously social and scientific senses. One way culminates in the dream of reduction to the *one* standard that trumps all others and then is raised to become the guiding principle of the societal and scientific order. This idea embodies the epitome of the standard per se, since the success of a homogenizing instrument is measured by the degree of unification it can achieve. But this dream bears a certain resemblance to that of the unified language that led to the Tower of Babel. It is a dream of the end of all conflicts and of the complete calculability of the incomparable uniqueness of each person. That's why it is ultimately a totalitarian dream—and so transforms into a phantasm.

The second path for standardization is gentler. It does not aim at the final leveling of all differences. Instead of striving for unification in space, it tries to create it through time. This happens not by means of a dream of an unchanging, eternal, and therefore nonnegotiable standard but by means of a factual implementation of technological instruments that can be rescinded—the introduction of reversibility. If standards are also always instruments of a scientific or societal consensus-finding, then the best standard is the one that can adapt effortlessly to the coincidences of the changing patterns of consensus. This standard carries within itself the option of revising itself, taking itself apart, and putting itself back together. Every negotiation can thus be turned around and reconsidered. The standards' reversibility arms them for future, unforeseeable imponderabilities, and their built-in flexibility guarantees their omnipresent (and omnifuture) use. This second option is the only one attractive to individuals in the molecular age who are ready to take their fate into their own hands. If pluralistic standards with built-in reversibility become the desired standard and thereby rule out a return and reduction to a single standard, then only making standards flexible enough can achieve a societal stabilization, however interim and precarious that stability may be. Flexibilization implies reversibility.

The dream of reversing time—reversing decisions that have already been made or reversing an entire life—is one of humanity's oldest myths, and it has been partially fulfilled by technology. It can be found in the craft of engineering, in law (as a human technology) with its "as if" fictions, and in biotechnology, where the biographical identity of a cell can be reprogrammed to revert it to the earliest phase of embryogenesis. On a tour of the Pergamon Museum in Berlin, we were shown how reversibility is practiced in archeological restorations. Since the finds from the past are usually fragments, they are put together in such a way that they can be taken apart and reassembled again at any time. Each reconstruction remains preliminary because new finds and more secure knowledge could turn up tomorrow. After our visit to the museum, we talked with the art historian Horst Bredekamp, who interprets this attitude as the sign of a "society of weak egos." In his view, people—including artists, politicians, and scientists—have grown risk-averse. No one has the courage any longer to carry out deeds that they are willing to stand by in the future. The ideal of perfection that inspired the Renaissance consisted in the artist's heroic gesture, in the grand plan, and in a will that included failure. The thought of submitting their ideal of perfection to a technology of reversibility would have seemed absurd and ridiculous to the artists of earlier times.

Yet some technological developments bear a high risk of failure. When the use of nuclear energy led to vehement controversies in the 1970s, the reversibility (and the failure-robustness in general) of technological solutions was also discussed. Considering the half lives of radioactive waste, one of the central questions is whether a society should permit itself to accept such long-term and apparently irreversible consequences. Today, a precautionary principle has entered into many states' environmental laws as a response to the unforeseeable human impact on the environment. It has embraced the concept of reversibility in that it stipulates that actions must be avoided whose consequences cannot be foreseen and that might be irreversible. And if nuclear energy is to be given a chance again, as currently seems possible, it will have to use technology to solve the problem of at least partial reversibility.

Engineers are familiar with the challenges of reversibility posed to technology. For science, too, reversibility is in principle nothing new, but here the problem arises in a different way. Reversibility means at least preliminarily immobilizing the arrow of time. For physics, that is no problem, but for biology, it is, because every living organism follows a life trajectory from birth to death. But the crucial question is *how* the organism follows this trajectory. Visions of immortality or of correspondingly postponed aging processes spur research powerfully, but the consensus among most scientists is that such goals lie in the distant future, at least for human organisms. Animal experiments show that, even in the molecular age, the most efficient method to slow down the aging process is to withhold calories, a method that has always been available and that our ancestors were probably aware of. Here, too, the "shock of the old" reveals deep continuities in the striving for knowledge and interventions that could be used to improve human life.

Modern biology's approach is much more in the direction of designing and developing organisms rather than seeking to make them immortal. The goal closest at hand, which is disturbing for many people, is to equip children with advantageous genetic variants rather than to protect them from aging. For this reason, reversibility in biology goes hand in hand with the paradigm of genetic circuits, with switching genes on and off, and with developing synthetic design, as we have repeatedly referred to. Genomes are manipulated in such a way that they can be set back to their original, untouched, and unmanipulated state. Synthetic modules are added as a supplement to be removed again if they should prove to be dangerous or useless. Reversibility is indispensable if biology—or at least a part of it—is to become a subdiscipline of the engineering sciences.

As we saw in the previous chapter, the sophisticated approach of altered nuclear transfer (ANT) can determine moral value by

suppressing a single gene and thus making cloning an ethically unproblematic process. ANT stands paradigmatically for the
principle of reversibility. Because we cannot be sure that the
gene whose removal blocks the full potential of development
and thereby answers moral objections will not later turn out to
be useful or even necessary for the stem cells gained from the
genetically crippled embryo, it is necessary that the process be reversible. The gene that was made inactive by means of a genetic
switch should be capable of reactivation with this same switch.
Some people see in this an incestuous connection between biotechnology and ethics, since the claimed moral value is to be
derived from the most contingent possible arrangement—an on/
off switch. But this switch is becoming an icon of the interwoven
development of science and societal values. The solution lies in
calibrated standards, and reversibility is their most salient trait.

To the degree that reversibility is introduced into societal
practices, we indeed approach the state that Bredekamp describes
as the "society of weak egos." By eschewing the heroic gesture
and the great, irrevocable attempt, reversibility has also left behind a past that made only a single future possible, one decided
in a present, creative moment, whereas for the society of weak
egos, the future can still be shaped in the future. It does not arise
in one single moment or in one attempt but piece by piece. It is a
patchwork, and thus it is open for the many minor tinkering interventions by people who want to contribute their experience,
including the mistakes they made in the past. A plural future
results that makes it possible to learn from mistakes and at least
partially revoke them. A society of weak egos can also swear off
the determinism implied by a monolithically conceived future.
It can approach a latently created future—a form of time that is
connected with indirect or postponed recognizability. Latency

always means more than simply absence. It is not some absence or other but, as Niklas Luhmann said, a "supporting absence."[7] A latent future permits something to arise that is open and accessible to human intervention. The result remains equally open. It depends on decisions (and on their consequences) that have not yet been made. Paradoxically, a society of weak egos is needed to encourage the individual to rely on his or her limited abilities and limited potential. The inherent tension between the individual and the collective that we have repeatedly encountered takes a surprising turn here.

SCOPE FOR THE CREATIVE INDIVIDUAL AND THE ACCOMPANYING INSTITUTIONS

Public campaigns for early diagnosis of breast cancer have given it greater visibility than other forms of cancer. The share that genetic factors have in the disease, which became widely known with the discovery of the BRCA1 and BRCA2 genes and the tests Myriad brought to the market, has created the figure of the "presymptomatic patient."[8] Empirical social research has investigated to what degree these presymptomatic tests play a pioneering role for genetic diagnoses and what difference it makes whether companies offer these tests on the free market, as in the United States, or whether health authorities manage demand by means of selection. It does not come as a surprise, therefore, that the media took up the story of a woman who decided to have her breast removed preventively before the appearance of symptoms.[9] She belonged to a genetically defined risk group, since close female relatives had already been affected by the disease. From our point of view, what is remarkable about this story is that the woman defined herself as a *previvor*. Here is a woman who made use of the entire repertoire offered by biomedicine to her in a standard-ized and consolidated form. She submitted to the correspond-ing tests, carefully weighed the probabilities, and subjected her decision for the preventive intervention to an equally carefully weighed cost-benefit analysis. At the end, the scales were tipped by her feelings. She consulted her diseased mother, who encour-aged her to stick with the decision she had already made.

In the statistics, this woman's decision, like those of many oth-ers, appears as evidence to be used for future methods of treat-ment. Her self-identification with a life to be lived preventively reaches us only as a story passed on by the media. In it, we en-counter a person who knows how to make use of the informa-tion she receives and who clearly takes her fate into her own hands. She changes her identity (the "feeling of identity," as Michael Pollak called it) in the face of a possible later affliction by declaring herself a preventive survivor. No one can say how much her decision was influenced by her surroundings or even by the medical establishment. But we do not hesitate to see her decision as a creative act.

We could adduce further examples of a subjective creative ap-proach to objective facts. We have already noted the empirical re-sults of the survey of identical twins who neither feel constrained by the existence of the other twin nor see in human cloning the horror scenario that this technology seems to evoke in a large part of the population. In anthropological studies, surrogate mothers and egg donors say their motivation was the desire to help others, even if the reasons differ. People affected by monogenetically in-herited Huntington's disease openly reveal their strategies about whether and when to pass on their knowledge to other members of their families and how they deal with a diagnosis for which no therapy exists.[10] Apparently, many who are affected are able to create an autonomous space for themselves and their stories in which they experiment with their lives and the heavy blows they receive. Gender indubitably still plays a role in this. But all these affected persons create meaning for something that, seen from outside, seems contingent and raises the unanswerable question: Why me? These persons order their social relationships anew and learn to deal with the uncertainties of a latent future that the standardized procedures of a superorganism reveal to them.

All in all, this increases the demand for preventive control. Robert Aronowitz has reconstructed the development of a cycle of fear and increased demand for prevention, using the example of the history of breast cancer in the United States.[11] In the experience of the affected persons, the disease and the risk of coming down with it become one and the same thing. The circle of the currently diseased is thereby expanded to include the potentially afflicted risk-bearers. The feeling of being able to do something oneself, even if the prospects of success may be low, increases the demand for preventive measures. The boundaries between symptomatic affliction, its preliminary phases, and their recognizability and thus the various risks of actually getting the disease are fluid. The risk of breast cancer thereby becomes the actual individual and societal health problem. It is equated with the disease that has fully broken out.

The missing link between the actor-centered and the system-oriented perspective is a political one—in the best sense of the term. Whether it is in a contemporary "politics of nature" (Bruno Latour), in the formation of the public realm around problem situations that the established institutions could not cope with (as John Dewey called for), or in other attempts to determine the common welfare, the collective space of experience in which individual experiences can be pluralistically bundled does not exist at present. To expect a return to common values shared by all is unrealistic, in our view. Whatever is just barely feasible in whatever is still common to a pluralistic life together must be found and tried out. This is why institutions are needed whose task is to test individual experiences for general validity, to learn from them, and to develop them further. The traditional, democratic political instruments are hardly adequate, and the newer instruments of a participatory and deliberative democracy soon reach their limits in the face of unbridgeable differences. It is true that a "counterdemocracy"[12] has given a voice to dissent and protest. Many demands, like those for more transparency and accountability, have been granted. But where are the principles of construction for designing institutions of a new type that can transform the individual space of expectation into a collective one?

One institution that approaches these framework conditions is the Human Fertilisation and Embryology Authority (HFEA) in Britain. It was founded in 1990 with the explicit goal of accompanying the societal and political adoption of new research possibilities that arose for the first time with the existence of human embryos in vitro. Since then, it has been active in all critical transitional phases that lead from the labs to the clinics and to important life decisions. The HFEA was active in an advisory and accompanying role in the rapidly expanding range of resulting research options—from the cloning of human embryos, through the use of preimplantation diagnosis for early detection of genetic variants associated with the risk of later disease (which includes screening embryos for BRCA1 and BRCA2 mutations), to the recently emerged possibility of inserting human cell nuclei into the enucleated oocytes of cows (interspecies cloning) to compensate for the rarity of human egg cells. This is not the place for a discussion of the advantages and limits of the individual decisions that the HFEA has made. But we would like to make special mention of two of the characteristics that enable the HFEA to accompany the biotechnological transformation and its selective adoption by society.

First is the HFEA's great representativeness. The supervisory board of the HFEA comprises representatives of a broad spectrum of British society—science, philosophy and ethics, religions, and consumer protection associations. The institution conducts regular public consultations as a kind of democratic experiment in which a theme is submitted comprehensively to the public. Focus groups and Web-based tools are thereby employed. The crux, however, is that the supervisory board does not claim to be completely representative and the result of the public consultations is not sufficiently legitimated in the classic democratic sense for the HFEA to take responsibility for it. For one thing, the members of the supervisory board are not elected. For another, if the consultations approve or reject an experimental procedure or a concrete application of research in reproductive medicine, this result is not simply accepted. This is not a referendum like those increasingly found in other European countries that try to involve the public in politically explosive themes resulting from advances in the life sciences. Referenda are coarse and extremely imprecise instruments. They reduce a complex, problematic matter to a simple yes-or-no question, where what is needed instead are nuanced, preliminary, and often only partial solutions.

If we view these two approaches from the perspective of the standards that we discussed in the previous chapter, then a referendum seeks to set up a single, final standard toward which everyone must orient himself or herself: something is either permitted or allowed. The HFEA's public consultations, by contrast, create spaces for constituting and comparing a number of competing, pluralistic standards. The point is not the citizens' private morals but a public morality that first emerges in argumentation in the process of consultations. Everyone can take part, including the organized groups that advocate universal, nonnegotiable standards, like claiming absolute protection for human embryos. But the HFEA does not automatically translate any of the positions, even if advocated by majorities, into a red or green light for the proposal under discussion. Instead, the proposals are subjected to a complex process of mediation in which the instruments of governance (as described in chapter 5) are employed. An individual citizen can turn to the HFEA to request a permit that would ease his or her personal situation (for example, as Diane Blood demanded the implantation of embryos conceived with the sperm of her deceased husband). Similarly, institutions can request a permit for specific research activities, as was the case with the transfer of human cells into bovine oocytes. Such requests are then also part of the complex process of deliberation. Majorities, individual voices, and positions advanced by laypersons and experts alike square off argumentatively in an ordered procedure that, despite vehement criticism of individual decisions, has won public credibility and trust.

The second conspicuous trait of HFEA governance is reversibility. In this regard, it operates firmly within the tradition of common law, which structures Britain's entire political system. Truths are not written down a priori and once and for all but arise fragmentarily in a process that distills them from individual cases—and that is precisely how they are further developed. Each HFEA decision about life options that are open to society or the individual is reversible by definition. But the mandate to review existing legislation at regular intervals is much more than a cautionary measure in the face of scientific and societal turbulence.

It reflects the primacy of a mature political system that has emancipated itself from the fiction of the natural.

In the 1990s, the British justice system was confronted with juridical struggles over the legal status of clones. Should a cloned embryo be treated analogously to one conceived "naturally" by IVF? The latter was an entity accepted by the British parliament in the 1990s when cloning was still unimaginable. Were clones merely another phase of life previously unknown to us, or was their conception so different that a new and public process of deliberation was needed to determine their ontological status? The answer ultimately came from the House of Lords. The Law Lords found that clones were similar enough to the embryos to be included in the same social order. But of greatest interest are the arguments underlying this decision:

If Parliament, however long ago, passed an Act applicable to dogs, it could not properly be applied to cats; but it could properly be applied to animals which were not regarded as dogs when the Act was passed but are so regarded now.[13]

According to this argument, dogs, cats, and whatever might lie between them are parts of an unfolding landscape in which what is natural is not given for all time but can unfold and develop further. One can ask who "sees" things this way, but the answer is obvious. The legislation of the country that brought forth empiricism—*our* view of things—and thereby strengthened our ability to change them, apparently regards itself as mature enough to take responsibility for the continuing naming of newly created forms of life. Here emerges the primacy of politics, which recognizes that in the molecular age the ideas of nature and the natural may be precise but are also preliminary and changeable approximations that remain open to human intervention. The molecular age creates a space between dogs and cats that a pluralistic *polis* is prepared to fill.

Despite all the objections made against it, we can at least see the contours of what an accompanying institution could be. The empowerment of the creative individual for an independent shaping of the future presupposes the greatest possible autonomy as well as material and nonmaterial resources. A society in which values and worldviews compete with each other under the omnipresent conditions of the market must extend its inventive capabilities to institutions that support individuals in selecting among options. The options available in an individual case cannot be imposed from above but require empathetic accompaniment of and respect for the individual's decision. Only then do the contours of a living together emerge in which one's own value system is not forced on others. Common experiences are not the sum of the experiences of all but a self-recognition in the experiences of others—and the ability, born from that self-recognition, to have compassion. This requires an accepted and respected institutionally secured consensus that people want to appropriate their multiple biotechnological future and integrate it in their lives in different ways. The era during which the *Mahabharata* was written and the molecular age lie far apart. But the questions posed back then about the condition of the era are still current today.

GLOSSARY

ALTERED NUCLEAR TRANSFER (ANT)   A recently proposed variation of somatic cell nuclear transfer in which a gene is mutated in the somatic cell nucleus prior to the nuclear transfer. The gene is chosen on the basis of its requirement for normal development.

BARCODE OF LIFE DATA SYSTEMS (BOLD)   An online project that supports the collection, management, analysis, and use of DNA barcodes. The project is based on the interspecies variability of short, standardized pieces of the genome. With animals, for example, researchers take a standardized gene region and assign all other comparable sequences to it. The goal is to take seawater or soil samples and identify known species in complex ecosystems or to discover new species.

BIOBANK   A collection of biological samples (such as gene fragments, entire genomes, cells, tissues, or organisms). It can be public or private or originate in a private-public partnership.

BLASTOCYST STAGE   One of the early stages in the development of the embryo. Embryonic stem cells are taken from embryos in this stage.

CELL NUCLEUS   In eukaryotic organisms, the part of the cell containing the vast majority of the genome.

CHIMERA   An organism that is made up of cells, tissues, or organs that derive from more than one organism. Thus, a transplant recipient is by definition a chimera. In this book, the term refers more narrowly to interspecies chimeras—organisms composed of cells, tissues, or organs that derive from organisms of different species.

CHORION   Greek, afterbirth. The outer membrane forming a sac around the embryo. The chorion villi sink into the mucous membrane of the uterus, forming the fetal part of the placenta through which material is exchanged with the mother.

CLONING   The process of isolating and propagating individual elements from various types of biological matter. Segments of DNA can be cloned, as can cells. In both cases, the key point is the ability to separate a specific item (whether a piece of DNA or a cell) from a more complex whole (the entire genome or an entire population of cells) and reproduce it at will. The cloning of DNA (also referred to as *recombinant DNA technology*) has been the foundation of molecular biology.

In this book, *cloning* refers to the technology employed to generate Dolly the sheep and a long list of other cloned animals. In this process, the nucleus is taken from a somatic cell. The nucleus contains most of a cell's DNA (called the *nuclear genome*), and the somatic cell can be any of the cells that make up the body (in distinction from the cells that comprise the germline eggs and sperm). This nucleus is inserted into an egg or a fertilized egg that has been deprived of its own nucleus. Development of the egg is triggered by artificial means (for example, an electric pulse), mimicking the process that occurs after fertilization, except that the resulting organism (embryo or fully grown animal) now has a nuclear genome identical to that of the somatic cell from which it was cloned. The mitochondrial DNA—an important but small portion of the cell's DNA that is contained in the energy-generating organelles of the cell (the mitochondria)—is provided by the host egg. This process is more precisely referred to as *somatic cell nuclear transfer*, since cloning (in analogy to DNA or cell cloning) implies the copying of the organism. As pointed out in chapter 1, however, copying life is a fiction, for both genetic and epigenetic reasons.

COMMON LAW   The law valid in many English-speaking countries that is based not on legislation but on authoritative past rulings by judges (precedents) and that develops accordingly.

COPRODUCTION   In science and technology studies (STS), the mutual influence between scientific and societal developments. The societal and scientific orders condition each other. Modern societies are characterized by science and technology that affect the living conditions of their individual members. Cultural, economic, and political prerequisites have to be fulfilled before certain scientific achievements and their spread in society are possible. For this reason, the organization of research, its epistemic goals, and its funding also change. Parallel developments are frequent, as is shown in chapter 7 in the example of standardization.

DEOXYRIBONUCLEIC ACID (DNA)   The full name of the polymer that makes up genes. It is a long chain of individual chemical entities known as *deoxyribonucleotides*.

EMBRYOGENESIS   In this book, the phase of an organism's development that takes place before birth.

EPIGENE, EPIGENETICS   From the Greek *epi* (above), the mechanisms of inheritance that are "above" the genetic. Cells that reproduce by means of cell division or entire organisms that reproduce by means of sexual or asexual reproduction transmit traits to the new generation. The inheritance of these traits entails a genetic component that is encoded in the actual sequence of the genes that are transmitted from cell to cell or from organism to organism. Many important traits, however, are inherited in an epigenetic fashion in the sense that their existence cannot be accounted for by the underlying DNA sequence. Furthermore, as is often the case for prominent terms in highly dynamic fields of research, the term *epigenetics* has taken on a life of its own, and in today's context it is often taken to mean the study of gene expression regulation and chromatin function.

GAMETES   Germ cells—oocytes and spermatozoa (egg and sperm cells) —that mediate the generation of new individuals through the inheritance of both genetic and epigenetic characteristics.

GENE   A functional unit made either of DNA or (for example, in the case of many viruses) of RNA. The concept has been robust because it is eminently flexible, encompassing genes that code for proteins (whose DNA sequence specifies the sequence of amino acids in the protein to be synthesized) as well as genes that function solely by virtue of being transcribed into molecules of RNA. Transcription is the process that generates an RNA molecule starting from the DNA molecule. In the process, the sequence stays the same (it is akin to a copying process), except that the building blocks used are slightly different—deoxyribonucleotides for DNA versus ribonucleotides for RNA.

GENE DOPING   Enhancing sport-relevant traits by altering genes in the cells of an athlete's body.

GENOME   The entire set of genes present in an individual cell or organism.

GENOTYPE   In the context of genetic analysis, the genetic component that underlies the inheritance of a given feature of interest (the phenotype).

GREEN FLUORESCENT PROTEIN (GFP)   A protein discovered in jellyfish and whose application in molecular biology was awarded the Nobel Prize for chemistry in 2008. Since its original identification, the gene encoding for GFP has become a central tool kit in every molecular biology laboratory. GFP emits fluorescent green light when exposed to UV rays or blue light and is routinely used to give visibility to proteins in living cells or to report the activity of specific genes in living organisms. GFP can be fused to a protein of interest to detect in living cells where this protein is normally localized. To locate in which tissues or organs a gene of interest is expressed, the gene can be replaced with GFP so that whenever that gene becomes active, the relevant cells, tissues, or organs will bear a green marker.

HETEROLOGOUS IN VITRO FERTILIZATION (IVF)   In vitro fertilization in which one or both gametes employed do not originate from the couple undergoing the IVF procedure. In the case of single women or men accessing IVF, the procedure is by definition heterologous.

HIGH-THROUGHPUT SEQUENCING   A recent improvement in sequencing technology that allows several rounds of sequencing to be performed in parallel, thus enabling the sequencing of entire genomes in a few weeks.

HUMAN FERTILISATION AND EMBRYOLOGY AUTHORITY (HFEA)   The institution established in 1990 in the United Kingdom by the Human Fertilisation and Embryology Act. Its mandate is to supervise both human-assisted reproduction and experimentation with human embryos. Clinics offering IVF and laboratories willing to perform research on human embryos both need to obtain a license from the HFEA.

HUMAN TECHNOLOGIES   Norms and rules whose function is to maintain and stabilize the social order. In analogy to material technologies, once they are introduced and accepted, they embody the achieved consensus.

Human technologies must be constantly adjusted to new requirements and to societal and technological changes.

INTERSPECIES CLONING   An experimental procedure recently approved in the United Kingdom to perform cloning research while circumventing the practical and ethical problems associated with the procurement of human eggs. The somatic cell nucleus that needs to be dedifferentiated or reprogrammed is transferred into an animal egg (usually cow or rabbit).

IUS, LEGAL PRACTICE   The activity of the judiciary, the courts, which results in the judicature of a specific legal question. It supplements legislation (*lex*).

LEX, LEGISLATION   The creation of legal norms. Stipulating legislative procedures is part of the minimum content of the constitution of any state. Lawmaking is the creation of legal norms and decisions. It is carried out by legislative or judicial bodies (*ius*).

METAGENOMICS   The attempt to register the entirety of the microorganisms in a genetic biotope. Unlike genomics, in which the DNA of a single organism is determined by means of high-throughput sequencing, in metagenomics genome sequences are taken from organisms that cannot be cultivated but that live together in their natural surroundings in a biotope. The goal of metagenomics is to gain insights into the structure of the biotope (biodiversity and distribution).

MITOCHONDRIAL DNA   The DNA that is present in the mitochondria, which represents a minute but important fraction of the cell's genome.

MODE 2 KNOWLEDGE   A form of knowledge production that is distinguished from the conventional disciplinary form of academic research (mode 1) in that new configurations of research groups from different disciplines and institutions work together on jointly defined problems. Mode 2 does not eliminate disciplinary boundaries, but it supplements the mode 1 work method with a flexible approach that transcends disciplinary and institutional boundaries and also provides more scope for social responsibility.

OOCYTES   The female germinal cells, commonly referred to as *eggs*.

PHENOTYPE   In the context of genetic analysis, the features of an organism that are assessed and followed through generations to determine their inheritance pattern.

POLYMERASE CHAIN REACTION (PCR) A technology that allows the in vitro amplification of specific DNA molecules, through the activity of an enzyme called *DNA polymerase*. At each step of the reaction, one segment of DNA is copied to make two, then each of these two to make two more, and so forth, resulting in an exponential increase in the number of copies of the original molecule. PCR was the first example of a hugely successful commercial application of biotechnology and remains to this day a cornerstone of laboratory practice throughout the world.

POSTGENOMICS The research that follows the complete sequencing of the genome of several organisms and that thereby aims at understanding the complex regulation that uses this genetic information in the organization of life forms.

PRECAUTIONARY PRINCIPLE A principle introduced in the context of environmental and health policy to ensure sustainability. According to it, damages and burdens must be avoided (or at least minimized) if possible, even if there is no complete scientific certainty about their effects. To apply it in practice, a comprehensive scientific evaluation should be carried out to determine the degree and kind of scientific uncertainty.

PREIMPLANTATION GENETIC DIAGNOSIS A diagnosis that is aimed at detecting genetic mutations in embryos before they are implanted in a woman's uterus.

PROCEDURALIZATION An observable trend in which the direct and final settlement of societal or legal conflicts is avoided in favor of defusing such conflicts by setting up procedural rules that are valid for all (and easier to follow).

PROTEINS Chains of amino acids encoded by genes. They are one of the main building blocks of living organisms. Their function is dependent on their three-dimensional structure. The sequence of the amino acids in each protein, which is the basic determinant of their three-dimensional structure, is determined by the sequence of nucleotides present in their encoding gene.

RECOMBINATION OF DNA (RDNA) The excision of specific fragments of DNA from the genome of organisms and their propagation (copying) in bacteria (and later also in other organisms) to characterize them on the level of sequence or function. This is the foundational technology of molecular biology.

RNA INTERFERENCE (RNAi) The inhibition of gene expression mediated by short molecules of double-stranded RNA that share a portion of the same sequence with the inhibited gene. Genes encode for proteins through RNA intermediates (called *messenger RNAs*). A gene is transcribed and spliced into a messenger RNA with a given sequence of nucleotides. The cell then uses this sequence as the template to assemble the corresponding protein in a process called translation. RNAi describes the mechanism whereby short molecules of double-stranded RNA target messenger RNA with the same sequence. This recognition can lead either to the destruction of the messenger RNA or to the inhibition of its translation into the relevant protein. In both cases, the outcome is a decrease in the degree to which the gene transcribed into the messenger RNA is expressed (used to generate the corresponding protein). The short molecules of double-stranded RNA can be administered to cells or organisms from the outside for experimental purposes. But recent years have revealed the key physiological relevance of this same pathway: cells generate short double-stranded RNA molecules through the cleavage of longer precursors and these endogenous double-stranded RNA molecules play essential roles throughout development and adult life.

SEQUENCING The determination of the sequence of nucleotides that make up a given segment of DNA or indeed an entire genome.

SOMATIC CELL See **cloning**.

SOMATIC CELL NUCLEAR TRANSFER See **cloning.**

SOMATIC CELL NUCLEUS See **cloning**.

STEM CELLS Cells that have two hallmark features pertaining to the types of cells that they can generate through cell division—self-renewal (the ability to give rise to more stem cells through cell division) and potential for differentiation (the ability to end self-renewal and give rise to daughter cells that will progressively differentiate into mature cell types). This dual property (the ability to self-renew while retaining the capacity to differentiate) constitutes the key feature of stemness.

STRUCTURAL GENETIC VARIATION　Variation among individual genomes that results from duplications, deletions, or inversions of entire stretches (hence structures) of nucleotides. Nucleotides are the building blocks of DNA commonly described by the four letters A, T, C, and G, which stand for the portion of their chemical structure that gives any DNA fragment its specific sequence. It is distinguished from the other and better-characterized type of genetic variation—single nucleotide polymorphisms (SNPs)—which is manifested by changes in individual nucleotides rather than in entire stretches of them.

SYNTHETIC BIOLOGY　Research aimed at characterizing, standardizing, and reassembling bits of biological matter into new configurations, systems, and life forms. Both in its theoretical underpinnings and in experimental practice, it is propelled by the ambition to turn biology into an extended domain of engineering. The minimal-genome project, which is attempting to define the smallest set of genes able to support life, is an example of the thrust of this line of research.

SYSTEMS BIOLOGY　Research aimed at understanding organisms as complex systems. The longstanding and still thriving tradition in molecular biology has been to focus on a specific gene or protein or on a small subset of them and probe their function in the organism. Large-scale approaches (the -omics approaches, such as genomics, proteomics, and metabolonics) have now made it possible to look at thousands of genes and proteins simultaneously. This evolution has paralleled the rise in bioinformatics approaches aimed at modeling the function of organisms by computing their entire sets of features (such as gene expression profiles) and inferring from them regulatory networks. Together, these developments are shifting the focus of inquiry to the comprehensive analysis of the features that underlie the dynamics of biological systems.

THERAPEUTIC CLONING　Using somatic cell nuclear transfer (see **cloning**) to derive from patients a tailored line of embryonic stem cells, which can be differentiated into the cells needed to replace damaged or diseased tissue. Since the cells have the same genome as the patient (except for the mitochondrial DNA), there is no risk of immune rejection after transplantation of the replacement cells. This approach has been successful in mice.

NOTES

I   THE VISIBILITY OF THE GENETIC FUTURE

1. Denis Diderot, *D'Alembert's Traum* (1769) (Leipzig: Reclam 1961), 15.

2. Alex Mauron, "Essays on Science and Society: Is the Genome the Secular Equivalent of the Soul?," *Science* 291, no. 5505 (2001): 831 f.

3. Thomas Nagel, *The View from Nowhere* (New York: Oxford University Press 1986).

4. Paul Rabinow, "Studies in the Anhropology of Reason," *Anthropology Today* 8 no. 5 (1992): 7–10.

5. Marilyn Strathern, *Kinship, Law and the Unexpected: Relatives Are Always a Surprise* (New York: Cambridge University Press 2005).

6. Hans Wigzell, "Science and Politics: When Ministers Are Well Primed," *Nature* 449, no. 7163 (2007): 663.

7. Barbara Prainsack and Tim D. Spector, "Twins: A Cloning Experience," in *Social Science and Medicine* 63, no. 10 (2006): 2739–2752; Barbara Prainsack, Lynn F. Cherkas, and Tim D. Spector, "Attitudes towards Human Reproductive Cloning, Assisted Reproduction and Gene Selection: A Survey of 4600 British Twins," *Human Reproduction* 22, no. 8 (2007): 2302–2308.

8. Bettina Bock von Wülfingen, *Genetisierung der Zeugung. Eine Diskurs- und Metaphernanalyse reproduktionsgenetischer Zukünfte* (Bielefeld: Transcript Verlag, 2007).

9. Staffan Müller-Wille and Hans-Jörg Rheinberger, *Heredity Produced: At the Crossroads of Biology, Politics, and Culture, 1500–1870* (Cambridge: MIT Press, 2007).

10. Hans-Jörg Rheinberger, *Experimentalsysteme und epistemische Dinge. Eine Geschichte der Proteinsynthese im Reagenzglas* (Göttingen: Wallstein Verlag, 2001).

2   THE GENETICIZATION OF ACHIEVEMENT

1. Paul Rabinow, *Marking Time: On the Anthropology of the Contemporary* (Princeton: Princeton University Press, 2008), 25 f.

2. Georges Canguilhem, "Nature dénaturée et nature naturante," in *Savoir, faire, espérer: Les limites de la raison,* Volume publié l'occasion du cinquantenaire de l'École des Sciences Philosophiques et Religieuses et en hommage à Mgr Henri Van Camp (Brussels: Publications des Facultés Universitaires Saint-Louis, 1976), 72–87.

3. Barbara Sahakian and Sharon Morein-Zamir, "Professor's Little Helper," *Nature* 450, no. 7173 (2007): 1157–1159.

4. Robert Gugutzer, "Die Fiktion des Natürlichen. Sportdoping in der reflexiven Moderne," *Soziale Welt* 52 (2001): 219–238.

5. Ivo van Hilvoorde, Rein Vos, and Guido de Wert, "Flopping, Klapping and Gene Doping: Dichotomies between 'Natural' and 'Artificial' in Elite Sport," *Social Studies of Science* 37, no. 2 (2007): 173–200.

6. Hans-Jörg Rheinberger, *Experimentalsysteme und epistemische Dinge. Eine Geschichte der Proteinsynthese im Reagenzglas* (Göttingen: Wallstein Verlag 2001).

7. Jennifer Henderson et al., "The EPAS1 Gene Influences the Aerobic-Anaerobic Contribution in Elite Endurance Athletes," *Human Genetics* 118, no. 3–4 (2005): 416–423.

8. C. H. Waddington, "Canalization of Development and Genetic Assimilation of Acquired Characters," *Nature* 183, no. 4676 (1959): 1654 f.

9. Lenny Moss, *What Genes Can't Do: Basic Bioethics* (Cambridge: MIT Press, 2003), 52.

10. Oliver W. Sacks, *Musicophilia: Tales of Music and the Brain* (New York: Knopf, 2007).

3   DISPUTED AFFILIATIONS: WHO BELONGS TO WHOM?

1. Skúli Sigurdsson, "Decoding Broken Promises," March 6, 2003, available at http://www.opendemocracy.net/theme_9-genes/article_1024.jsp (accessed on May 12, 2010).

2. Arthur A. Bergen et al., "Mutations in ABCC6 Cause Pseudoxanthoma Elasticum," *Nature Genetics* 25, no. 2 (2000): 228–231.

3. "The Advocates" (editorial), *Nature Genetics* 38 (2006): 391.

4. Shobita Parthasarathy, *Building Genetic Medicine: Breast Cancer, Technology, and the Comparative Politics of Health Care*, Inside Technology Series (Cambridge: MIT Press, 2007).

5. Samuel Levy et al., "The Diploid Genome Sequence of an Individual Human," *PLoS Biology* 10 (2007): 254.

6. David A. Vise and Mark Malseed, *Die Google-Story* (Hamburg: Murmann, 2007), 268–276.

7. Ibid., 286.

4   CONTESTED FUTURES: EVERYDAY EXPERIENCE AND THE VALUES DISCOURSE

1. Gaston Bachelard, *Die Bildung des wissenschaftlichen Geistes. Beiträge zu einer Psychoanalyse der objektiven Erkenntnis* (Frankfurt am Main: Suhrkamp 1987).

2. Wolf Lepenies, "Vergangenheit und Zukunft der Wissenschaftsgeschichte—Das Werk Gaston Bachelards," in Bachelard, *Die Bildung des wissenschaftlichen Geistes*, 7–34.

3. Harry M. Collins and Robert Evans, *Rethinking Expertise* (Chicago: University of Chicago Press 2007).

4. Charles Taylor, *Sources of the Self: The Making of the Modern Identity* (Cambridge: Harvard University Press, 2007); Charles Taylor, *A Secular Age* (Cambridge: Belknap Press of Harvard University Press, 2007).

5. David A. Vise and Mark Malseed, *Die Google-Story* (Hamburg: Murmann, 2007), 268–276.

6. Sarah Franklin, "The Cyborg Embryo," *Theory, Culture and Society* 23, no. 7–8 (2006): 167–187.

7. Pierre Hadot, *The Veil of Isis: An Essay on the History of the Idea of Nature* (Cambridge: Belknap Press of Harvard University Press, 2006).

8. Keisuke Okita, Tomoko Ichisaka, and Shinya Yamanaka, "Generation of Germline-Competent Induced Pluripotent Stem Cells," *Nature* 448, no. 7151 (2007): 313–317.

9. Dani le Hervieu-Léger, "The Role of Religion in Establishing Social Cohesion," in *Religion in the New Europe*, ed. K. Michalski (New York: Central European University Press, 2006), 45–63.

10. Nikolas Rose, *The Politics of Life Itself: Biomedicine, Power, and Subjectivity in the Twenty-First Century* (Princeton: Princeton University Press, 2006).

11. John Dewey, *The Public and Its Problems* (Athens: Swallow Press, 1991).

12. Walter Lippmann, *The Phantom Public,* The Library of Conservative Thought (New Brunswick: Transaction Publishers, 1993; 1st edition 1925), 121.

13. Stein Ringen, *The Liberal Vision and Other Essays on Democracy and Progress* (Oxford: Bardwell Press, 2007).

5 CONTESTED FUTURES: EVERYDAY EXPERIENCE AND THE VALUES DISCOURSE

1. Maurice Godelier, *Au fondement des sociétés humaines: Ce que nous apprend l'anthropologie* (Paris: Bibliothèque Albin Michel, 2007), 65 (quotation here translated by Mitch Cohen).

2. Michael Critchon, *Next: A Novel* (New York: Harper Collins, 2006).

3. Alain Supiot, Homo juridicus: *Essai sur le fonction anthropologique du droit* (Paris: Seuil, 2005), Homo juridicus: *On the Anthropological Function of the Law*, trans. Saskia Brown (London: Verso, 2007.

4. Sheila Jasanoff, *Designs on Nature: Science and Democracy in Europe and the United States* (Princeton: Princeton University Press, 2005).

5. Freeman Dyson, "Our Biotech Future," *The New York Review of Books* 54, no. 12 (2007): 4–8.

6. Supiot, Homo juridicus.

7. Renate Mayntz, "Governance im modernen Staat," in Arthur Benz, *Governance—Regieren in komplexen Regelsystemen. Eine Einführung* (Wiesbaden: VS Verlag für Sozialwissenschaften, 2004), 65–76.

8. Ibid.

9. Herbert Gottweis, "Governing Genomics in the Twenty-first Century: Between Risk and Uncertainty," *New Genetics and Society* 24, no. 2 (2005): 175–193.

10. John Hyde Evans, *Playing God? Human Genetic Engineering and the Rationalization of Public Bioethical Debate* (Chicago: University of Chicago Press, 2002).

11. Brian and Charlotte Salter, "Bioethics and the Global Moral Economy: The Cultural Politics of Human Embryonic Stem Cell Science," *Science, Technology and Human Values* 32, no. 5 (2007): 554–581.

12. Ibid.

13. Charles L. Bosk, "Logical Professional Ethicist Available: Logical, Secular, Friendly," *Daedalus* 128, no. 4 (1999): 47–68.

14. John Durant, Martin Bauer, and George Gaskell, eds., *Biotechnology in the Public Sphere: A European Sourcebook* (London: Science Museum, 1998), 226.

6   SYNTHESIZING THE FUTURE

1. W. Szybalski and A. Skalka, "Nobel Prizes and Restriction Enzymes," *Gene* 4, no. 3 (1978): 181 f.

2. Thomas F. Knight, cited in Drew Endy, "Foundations for Engineering Biology," *Nature* 438, no. 7067 (2005): 449–453.

3. Shibu Yooseph et al., "The Sorcerer II Global Ocean Sampling Expedition: Expanding the Universe of Protein Families," *PLoS Biology* 5, no. 3 (2007): 16.

4. Sylvie Coyaud, "Nei taccuini il romanzo della vita," *II Sole-24 Ore* (February 10, 2008): 40.

5. http://parts.mit.edu/registry/index.php/Main_Page.

6. Dinah Birch, "A Brief History of the Future," *Times Literary Supplement*, February 1, 2008, 19, review of Brian Aldiss, ed., *A Science Fiction Omnibus* (London: Penguin, 2007).

7. Keisuke Okita, Tomoko Ichisaka, and Shinya Yamanaka, "Generation of Germline-Competent Induced Pluripotent Stem Cells," *Nature* 448, no. 7151 (2007): 313–317.

8. William B. Hurlbut, "Altered Nuclear Transfer as a Morally Acceptable Means for the Procurement of Human Embryonic Stem Cells," *National Catholic Bioethics Quarterly* 5, no. 1 (2005): 145–151.

9. Katrien Devolder, "What's in a Name? Embryos, Entities, and ANTities in the Stem Cell Debate," *Journal of Medical Ethics* 32, no. 1 (2006): 43–48.

10. Hurlbut, "Altered Nuclear Transfer."

11. Alexander Meissner and Rudolf Jaenisch, "Generation of Nuclear Transfer-Derived Pluripotent ES Cells from Cloned Cdx2-Deficient Blastocysts," *Nature* 439, no. 7073 (2006): 215.

12. Davor Solter, "Politically Correct Human Embryonic Stem Cells?," *New England Journal of Medicine* 353, no. 22 (2005): 2321–2323.

7   THE FUTURE AS SYNTHESIS

1. James A. Byrne et al., "Producing Primate Embryonic Stem Cells by Somatic Cell Nuclear Transfer," *Nature* 450 (2007): 497–502.

2. F. Jacob, *Die innere Statue. Autobiographie des Genbiologen und Nobelpreisträgers* (Zurich: Ammann; Hans-Jörg Rheinberger, 1988); F. Jacob, *Experimentalsysteme und epistemische Dinge. Eine Geschichte der Proteinsynthese im Reagenzglas* (Göttingen: Wallstein Verlag, 2001), 22.

3. Geoffrey C. Bowker and Susan Leigh Star, *Sorting Things Out: Classification and Its Consequences (Inside Technology)* (Cambridge: MIT Press, 1999).

4. Virginie Tournay, *La gouvernance des innovations médicales* (Paris: Presses universitaires de France, 2007).

5. Helga Nowotny, Peter Scott, and Michael Gibbons, *Re-thinking Science* (Oxford: Polity, 2001).

6. Lee M. Silver, *Challenging Nature: The Clash of Science and Spirituality at the New Frontiers of Life* (New York: Ecco, 2006).

7. Stefanie Diekmann and Thomas Khurana, *Latenz* (Berlin: Kulturverlag Kadmos, 2007), 12.

8. Monica Konrad, *Narrating the New Predictive Genetics: Ethics, Ethnography, and Science,* Cambridge Studies in Society and the Life Sciences (Cambridge: Cambridge University Press, 2005), 1.

9. Amy Harmon, "Free of Breast Cancer, but Weighing a Mastectomy Because of Genetic Tests," *International Herald Tribune*, September 17, 2007, available at http://www.finzfirm.com/diseases/Breast_Cancer/Free_of_breast_cancer_but_weighing_a_mastectomy_because_of_genetic_tests/ (accessed May 12, 2010).

10. Konrad, *Narrating.*

11. Robert A. Aronowitz, *Unnatural History: Breast Cancer and American Society* (New York: Cambridge University Press, 2007).

12. Pierre Rosanvallon, *La contre-démocratie: La politique à l'âge de la défiance* (Paris: Seuil, 2006).

13. Opinion of Lord Bingham, UKHL 13, para. 15, 2003, cited by Sheila Jasanoff, *Designs on Nature: Science and Democracy in Europe and the United States* (Princeton: Princeton University Press, 2005), 200.